~ THAI ~
MASTERCLASS
COOKBOOK

Simple Thai Recipes

With a Twist

SOMCHAI THEERAVIT

Table of Contents

PART I

Chapter 1: Noodles - Rice & Soup

Drunkard's Noodles

Servings Provided: 1-2

Time Required: 15 minutes

What is Needed:

- Wide rice noodles- prepared dry noodles (About 2 cups)
- Chicken meat/pork loin (.5 cup - chicken breast)
- Baby corn (1-2)
- Carrots (2 -3 small)
- Thai long chili - mild - yellow/red/orange (2-3 thin slivers)
- Fresh green peppercorns (1 bunch - leave the corns in the bunch)
- Finger root (10 tiny flakes/one average finger - cut into slivers)
- Thai hot chili (3-4 - whole red & green)
- Brown/yellow onion (half a slice)
- Garlic - not peeled (3-4 cloves)
- Thai Sweet Basil/Star of siam basil (1 cup)

- Dark sweet soy sauce (.5 tsp.)
- Oyster sauce (2 tsp.)
- White cane sugar (1 tbsp.)
- Light soy sauce (2 tsp.)

Preparation Method:

1. Fry the garlic, chili, and onions in oil for about one minute.
2. Cut the carrot lengthwise into slices. Mix in the corn, carrots, sweet chili, and green peppercorn bunch. Stir and simmer for another 10-15 seconds.
3. Slice and mix in the chicken and cook until nearly done, and add the soy sauce, sugar, and oyster sauce. Simmer until the sugar dissolves.
4. Add the pre-cooked rice noodles, coating them with the sauces.
5. Transfer the pan to a cool burner and garnish it with the fresh basil. Serve and enjoy it promptly.

Jasmine Rice

Servings Provided: 4-6

Time Required: 20 minutes

What is Needed:

- Water (2.75 cups + as needed)
- Jasmine rice (1.5 cups)
- Salt (.75 tsp.)

Preparation Method:

1. Boil the water with the salt. Stir in the rice, put a lid on the pot, and adjust the temperature setting to low. Let it simmer until all of the water is absorbed (15 min.).
2. Taste test the rice. Add a few tablespoons of water if it's too firm. Cover the pan and let the rice absorb the water.

Chicken Wonton Soup - Kiao Nam Gai

Servings Provided: 6-10

Time Required: Less than 45 minutes

What is Needed:

The Wontons:

- Ground chicken or pork (400 grams/about 14 oz.)
- Lemongrass (1 tsp.)
- Kaffir lime leaves (1 tsp.)
- Coriander Root (1 tsp.)
- Green onion (1 tsp.)
- Garlic (1 tsp.)
- Tapioca Starch (1 tsp.)
- White pepper - finely ground (1 tsp.)
- Sesame oil (.5 tsp.)

- Fish Sauce (.5 tsp.)
- Wonton wrappers (50)
  The Broth:

- Chicken or vegetable stock (6-8 cups)
- Coriander root (1 tsp.)
- Garlic (1 tsp.)
- Oyster sauce (2 tsp.)
- Light soy sauce (1 tsp.)
- Sugar (1 tsp.)
- White pepper (.5 tsp.)
- Chinese cabbage/Bok Choy Leaves *(4 - 6 heads)*

Preparation Method:

1. Grind the kaffir lime leaf, lemongrass, garlic, and coriander root into a smooth paste. Blend the mixture with the chopped meat, tapioca flour, green onion, white pepper, fish sauce, and sesame oil.
2. Fill and wrap the wontons.
3. Prepare the soup broth by adding white pepper, garlic, coriander root, oyster sauce, and soy sauce to the unseasoned chicken or vegetable stock. Boil hard for about five minutes.
4. In another soup pot, warm water and boil the wontons for about five minutes. Remove them and place them in a serving bowl.
5. Blanch the bok choy leaves for about one minute and place them in your bowls with the wontons.
6. Spoon enough of the broth over the bowl to cover the wontons and bok choy lightly.
7. Garnish them using chopped green onions and a bit of white pepper.
8. Serve with condiments, including sun-dried pickled green chilies, red chili flakes, sugar, and fish sauce with red chili.

Herbal Chicken Soup With Bitter Melon

Servings Provided: 4

Time Required: Less than 15 minutes

What is Needed:

- Breast portion chicken (half of 1)
- Bitter melon/Chinese bitter gourd (half of 1)
- Cilantro/Coriander leaves - roots attached (3-4)
- Galangal root (1-2-inch piece)
- Lemongrass (1 stalk)
- Thai hot chilies (2 or more)
- Garlic (2-3 cloves)
- Sugar (1 tsp.)
- Light soy sauce (1 tbsp.)
- Coconut/vegetable oil (1 tbsp.)

Preparation Method:

1. Discard the skin and chop the chicken into small cubes. Remove the seeds and inner pulp from the bitter melon and cut it also.
2. Dice the coriander and grind the garlic, red chili, and galangal into a rough pulp.
3. Prepare a skillet of coconut oil to fry the herbs for about half a minute and add the chicken to fry until cooked - slowly.
4. Pour in water and wait for it to boil.
5. Add the diced melon and simmer until tender (8 min.). Extinguish the heat and toss in the chopped coriander leaves.
6. Serve the soup while it's hot with a sprinkle of chopped coriander just before serving.

Pumpkin Coconut Soup

Servings Provided: 2-3

Time Required: 25 minutes

What is Needed:

- Chicken stock (6 cups)
- Freshly minced lemongrass (4 tbsp.)
- Makrut lime leaves (3 left whole) or substitute Lime zest(1 tsp.)
- Shallot (1 minced) or Purple onion (.25 cup minced)
- Garlic (3 cloves)
- Galangal or ginger (1 thumb-size piece)
- Fresh red chili (1) or dried crushed chili (.25-.5 tsp.) or chili sauce (1-2 tsp.)
- Pumpkin/squash (3 cups)
- Sweet potato/yams (2 cups)
- Ground coriander (.75 tsp.)
- Turmeric (.5 tsp.)

- Ground cumin (1 tsp.)
- Fish sauce (2 to 3 tbsp./as desired)
- Shrimp paste (.5-1 tsp.) or fish sauce(1 extra tbsp.)
- Brown sugar (1 tsp.)
- Lime juice - fresh-squeezed (2 tbsp.)
- Thick coconut milk (1/3 to 1/2 can)
- Soft tofu (1 to 2 cups - sliced into cubes/substitute chickpeas or cooked shrimp)
- Baby spinach - washed (1 generous handful)
- Fresh basil - the topping (.5 cup)

Preparation Method:

1. Pour the stock into the pot to warm using the high-temperature setting.
2. Rinse the spinach.
3. Mince and add the lemongrass (the stalk too if it's fresh), garlic, shallot, the makrut lime leaves, galangal or ginger, and chili. Wait for it to boil.
4. Peel, dice, and mix in the squash and yam. Adjust the temperature setting slightly and gently boil for six to seven minutes.
5. Meanwhile, whisk the spices and stir well after each addition (ground coriander, turmeric, cumin, shrimp paste, fish sauce, lime juice, and brown sugar).
6. Once the pumpkin and yams are softened, adjust the temperature setting to low. Fold in the coconut milk as desired.
7. Give it a taste test to adjust the seasonings to your liking.
8. Just before serving, gently mix in the spinach and softened tofu.
9. Serve and garnish it using the coriander and basil.
10. Enjoy it with a serving of noodles or rice.

Red Curry With Bamboo Shoots & Coconut Milk

Servings Provided: 2

Time Required: 20 minutes

What is Needed:

- Boneless - skinless chicken thigh meat (1 lb./cut into ½-inch pieces)
- Coconut milk (2 cups - divided into two portions)
- Thai red curry paste (3 tbsp.)
- Bamboo shoots - canned or prepared - cut into thin strips (1 cup)
- Thai long chilies/another mild red chili (2-3)
- Thai sweet basil/Horapah-Star of Siam (1 cup)
- Kaffir lime leaves (5)
- Coconut Sugar (2 tsp.)
- Salt (.5 tsp./as needed)
- Fish Sauce (1 tbsp./as needed)

Preparation Method:

1. To get started, warm half the coconut milk in a wok or pan. Stir in the curry. Cook it using the low-temperature setting until it is thick and well blended.
2. Add the chicken stir-fry in the curry for about five minutes, and add the other half of the coconut milk.
3. Fold in the bamboo shoots, pieces of kaffir lime leaf, and red chilies. Simmer it for five minutes and let it cool.
4. Stir in the basil and serve

Chapter 2: Poultry

Cashew Chicken

Servings Provided: 2-4

Time Required: 20 minutes

What is Needed:

- Chicken legs - boneless & skinless (1 lb)
- Unsalted - raw cashew nuts (.5 cup)
- Thai long chilies (red & green/4-6)
- Wax peppers (1-2)

- Sun-dried Thai long chili (3-4)
- Brown/yellow onion (half of 1 small)
- Green onions (2-3)
- Garlic (2-3 large cloves)
- A-P flour (.25 cup)
- Finely cracked white pepper (.25 tsp.)
- Dark sweet soy sauce (.5 tsp.)
- Oyster sauce (2 tbsp.)
- Sugar (1 tsp.)
- Light soy sauce (.5 tsp.)
- For Frying: Vegetable Oil (1 cup)
- Water (.25 cup/as needed)

Preparation Method:

1. Dice the chicken into chunky pieces. Dice the onion, chilies, and green onions, and garlic.
2. Deep-fry the dried chilis, cashew nuts, and the chicken. Set them aside and dump the oil (reserving one tablespoon).
3. Sauté the garlic and onion. Mix in the sauces and sugar with a bit of water to make a gravy.
4. Fold in the fried chicken and stir until coated in the gravy. Simmer it for about two minutes.
5. Add the fried cashew nuts and simmer for a minute, and add the fresh chilies. Thoroughly mix and cook for 30 seconds.
6. Serve topped with the fried, dried chili pieces.

Chicken & Basil Stir-Fry

Servings Provided: 3-4

Time Required: minutes

What is Needed:

- Bone-in with skin chicken breast halves (1 lb.)

- Vegetable oil (1 tbsp.)
- Fresh ginger & garlic (1 tbsp. each)
- Hot chili flakes (.25 tsp.)
- Fat-skimmed chicken broth (2/3 cup)
- Soy/Asian fish sauce (1 tbsp.)
- Cornstarch (2 tsp.)
- Fresh basil leaves (3 cups - lightly packed)
- Salt
- Also Needed: Nonstick skillet (8-10-inch)

Preparation Method:

1. Mince the garlic and ginger.
2. Rinse and dab the chicken dry. Slice it crossways into strips 2-3 inches long and are about 1/8-inches thick.
3. Warm the skillet using the high-temperature setting and add the oil.
4. Toss in the chicken, chili flakes, ginger, and garlic.
5. Stir frequently and cook for about three to four minutes. (The pink in the center should be gone.
6. Whisk the cornstarch, fish sauce, and broth in a mixing container. Pour it into the pan and stir for about one minute.
7. Mix in the basil leaves, stirring about 30 seconds until they are wilted and dust with salt to serve.

Easy Thai Chicken

Servings Provided: 8

Time Required: 40 minutes

What is Needed:

- Unsalted butter (2 tbsp.)
- Peanuts (.25 cup)
- Cilantro leaves (2 tbsp.)
- Chicken thighs (8)

The Sauce:

- Sweet chili sauce (.5 cup)
- Soy sauce (2 tbsp.)
- Garlic (2 cloves)
- Fish sauce (1 tbsp.)
- Fresh ginger (1 tbsp.)
- Sriracha (1 tsp./as desired)
- Juice of 1 lime

Preparation Method:

1. Set the oven temperature at 400° Fahrenheit.
2. Chop the cilantro leaves and peanuts. Mince/grate the garlic and ginger.
3. Prepare the sauce by whisking the soy sauce, fish sauce, chili sauce, ginger, garlic, Sriracha, and lime juice in a mixing container and set it aside.
4. Prepare a large oven-proof skillet using the med-high temperature setting to melt the butter.
5. Add the chicken to the skillet and sear it (skin-side down) until golden brown (2-3 min. per side). Stir in chili sauce mixture.
6. Set a timer and roast until it reaches an internal temperature of 165° Fahrenheit (25-30 min.).
7. Switch it to the oven to broil to cook until caramelized and slightly charred (2-3 min.).
8. Serve them promptly topped with cilantro and peanuts.
9. Note: It's okay to leave the skin and bones on the chicken.

Ginger Chicken

Servings Provided: 2-4

Time Required: 20 minutes

What is Needed:

- Chicken breast (2 cups)
- Fresh ginger (1 cup)
- Cloud Ear Fungus (4 oz./about 100gm)
- Green onions (2 to garnish & 2 for the recipe/4 whole)
- Thai long chili/another mild red chili (2)
- Garlic (3-4 large cloves)
- Oyster sauce (2 tbsp.)

- Light soy sauce (2 tsp.)
- Palm/light brown sugar (1 tsp.)
- Black pepper (.5 tsp.)
- Vegetable oil (1 tbsp.)

Preparation Method:

1. Do the prep. Thinly slice the chicken. Peel and julienne the ginger and mince the garlic.
2. Sauté the garlic in the oil for about half a minute, and arrange the chicken in the skillet. Sauté it for about two minutes, stirring continuously.
3. When the chicken is almost done, add the sugar, black pepper, soy sauce, and oyster sauce. Stir and fry until the sugar melts, and the pan is sticky. Add a little water to make a thick sauce.
4. Add the Mouse Ear fungus and cook for another two minutes.
5. Fold in the ginger, green onion, and red chili.
6. Toss all of the fixings and continue to stir-fry for about one minute until it's all piping hot. Serve promptly.

Green Curry Chicken

Servings Provided: 4

Time Required: 28 minutes

What is Needed:

- Green curry paste (2 tbsp.)
- Fresh ginger (1 tbsp.)
- Light coconut milk (14 oz. can)
- Fish sauce (2 tsp.)
- Shredded dark meat chicken (.75 cup)
- Shredded chicken breast (.5 cup)
- Lime juice (1 tbsp.)
- Garlic clove (1 minced)
- Sesame oil (1 tbsp.)
- Sliced baby bok choy (3 cups)
- Sliced red bell pepper (1 cup)
- Uncooked wide brown rice noodles (5 oz.)
- Cilantro leaves (.5 cup)
- Lime (4 wedges)

Preparation Method:

1. Peel and grate the ginger. Combine the first four fixings (up to the
   line) in a saucepan using the medium-high temperature setting. Once
   boiling, adjust the setting to simmer for 15 minutes. Fold in the

chicken and simmer for another five minutes.

2. Transfer the pan to a cool burner and mix in the freshly-squeezed lime juice.

3. Warm oil in a skillet using the med-high temperature setting. Mix in the bok choy and bell pepper. Sauté them for two minutes and mix in the garlic. Continue to sauté them for 30 seconds.

4. Prepare the noodles per the package direction. Dump it into a colander to drain. Portion the noodles into four servings, topping each with ½ cup chicken mixture and ½ cup of the vegetables.

5. Garnish the chicken using the cilantro, and serve with a side of lime wedges.

Spicy & Salty Fried Chicken Wings

Servings Provided: 8-12

Time Required: 30 minutes

What is Needed:

- Chicken wings (3 lb.)
- Thai red curry paste (2 tbsp./as desired)
- Ground - dried Thai hot chili (1-2 tsp.)
- Oyster sauce (1 tbsp.)
- Sugar (1 tsp.)
- Fish sauce (1-2 tsp.)

- Water (.5 cup as needed)
- Optional Garnish: Kaffir lime leaves (6-8)
- Vegetable oil (.5 cup/as needed)

Preparation Method:

1. Whisk the fish sauce, oyster sauce, sugar, ground chili, and one tablespoon of red curry with the water.
2. Pour the mixture over the wings to marinate for a minimum of four hours.
3. Transfer the marinated chicken to a platter.
4. Mix in about one to two tablespoons of flour to the sauce and whisk it to create a thin batter.
5. Dust the chicken with flour, and dip them into the batter until they're well coated.
6. Deep fry them (in oil) for three to five minutes per side. Place them on a layer of paper towel towels to drain the excess fat.
7. Top with a portion of the crispy fried kaffir lime leaves.

Spicy Stir-Fried Chicken With Eggplant

Servings Provided: 2-4

Time Required: 10 minutes

What is Needed:

- Whole chicken leg (1 large - boneless & skinless)
- Thai eggplants (6-8)
- Thai long chilies/Mild green chiles (3-4)
- Thai sweet basil/leaves & flower tops (1 cup)
- Light soy sauce (2 tsp.)
- Sugar (.5 tsp.)
- Thai red curry paste (1.5-2 tbsp.)
- Fish sauce (2 tsp.)

Preparation Method:

1. Chop the chicken into bite-sized pieces. Pluck the leaves from the basil and cut the eggplant and chilis.
2. Fry the chicken with the red chili until thoroughly cooked. Add in the eggplant and continue cooking for about three to four minutes.
3. Mix in the mild green chili, and simmer it for another minute, and transfer the pan from the burner.
4. Garnish it using basil and enjoy the dish promptly.

Chapter 3: Beef

Basil Beef

Servings Provided: 4

Time Required: 30 minutes

What is Needed:

- Vegetable oil (2 tbsp.)
- Shallots (2)
- Garlic (7 cloves)
- Fresh ginger (1 tbsp.)
- Red bell pepper (half of 1)
- Lean ground beef (1 lb.)
- Brown sugar (2 tsp.)
- Soy sauce - low-sodium (6 tbsp.)
- Fish sauce (2 tbsp.)
- Asian garlic chili paste (2 tbsp./to taste)
- Oyster sauce (3 tsp.)
- Beef broth - low-sodium (.5 cup)
- Water (.25 cup)
- Cornstarch (1 tsp.)
- Basil leaves (1 cup)
- For Serving: Jasmine rice

Preparation Method:

1. Thinly slice the shallots and pepper. Mince the ginger and garlic. Set them aside.

2. Warm a large skillet/wok using the med-high temperature setting to heat the oil. When it's hot, toss in the peppers, ginger, garlic, and shallots to stir-fry for about three minutes.

3. Push the vegetables to the side and adjust the temperature setting to high. Add the beef, breaking it apart as it cooks.

4. Whisk the water, broth, cornstarch, chili paste, oyster sauce, fish sauce, soy sauce, and brown sugar. Simmer the sauce for about two minutes.

5. Toss in the basil and stir fry until it's wilted. Serve over cooked rice.

Steak Noodle Bowl

Servings Provided: 4

Time Required: 21 minutes

What is Needed:

- Green cabbage (2.5 cups)
- Freshly squeezed lime juice - divided (1 tbsp.)
- Kosher salt (1 tsp. - divided)
- Sugar (2 tsp. - divided)
- Uncooked flat-brown rice noodles - ex. Annie Chun's pad Thai-type (8 oz.)
- Top sirloin steak (12 oz.)
- Canola oil (1.5 tsp.)
- Water (.5 cup)
- Red curry paste - ex. Thai Kitchen (2 tbsp.)
- Light coconut milk (13.5 oz. can)
- Optional: Lime wedges (4)

Preparation Method:

1. Thinly slice the cabbage and add one teaspoon of juice, ½ teaspoon sugar, and ¼ teaspoon of salt. Toss and set it aside for about 15 minutes.
2. Prepare the noodles per the instructions provided on the package. Dump them into a colander to rinse and drain using fresh-cool water.

3. Toss the steak with about ½ teaspoon of sugar.

4. Prepare a skillet using the high-temperature setting to warm the oil.

5. Trim and slice the steak into thin strips and add it to the pan to simmer for about two minutes. Flip it over to cook for another half minute and remove it from the pan. Set it aside and keep it warm.

6. Add about ½ of a cup of water to the pan, scraping it to loosen browned bits. Add the curry paste and coconut milk, stirring well to combine. Wait for it to simmer.

7. Adjust the temperature setting to low and continue cooking for five minutes. Mix in the remaining two teaspoons lime juice, remaining one teaspoon sugar, and ½ teaspoon salt.

8. Arrange about one cup of the noodles in each of the four bowls and add the steak.

9. Ladle about ½ cup of the broth over each serving and top each with about ½ cup of the cabbage mixture. Sprinkle the remaining ¼ teaspoon of salt evenly over each of the servings. Serve with lime wedges, if desired.

Chapter 4: Seafood

Thai Fish Cakes

Servings Provided: 20 small cakes

Time Required: 25 minutes

What is Needed:

- Whitefish fillets - cod or halibut (approx.13.5 oz.)
- Red curry paste (2 tbsp.)
- Green beans (.75 cup)
- Egg (1)

- Cornstarch (1 tbsp.)
- Sugar (1 tsp.)
- Fish sauce (1 tbsp.)
- Lime leaves (5)
- Vegetable oil (3 cups)

Preparation Method:

1. Chop the fish into small pieces and toss it into a food processor.
2. Break and add in the egg, cornstarch, curry paste, sugar, fish sauce, and blend until mixed. (*Don't puree* the mixture.)
3. Finely slice the green beans and lime leaves to combine with the mixture.
4. Form the mixture into patties, using about one tablespoon and shape until they are about two inches in diameter.
5. Add oil to a saucepan using the high-temperature setting and deep-fry the cakes.
6. Once the oil is sizzling, arrange the cakes in the saucepan and cook them until they are golden brown. Place the fish cakes onto a layer of paper towels to absorb the oil before serving.
7. Serve hot with sweet chili sauce (see the recipe below) with cucumber and crushed peanuts.

Sweet and Chili Sauce

What is Needed:

- Sugar (3 tbsp.)
- White vinegar (2 tbsp.)
- Salt (.25 tsp.)
- Water (2 tbsp.)
- Garlic (1 clove)
- Fresh chilies (2)
- Cucumber (2 tbsp.)
- Crushed peanuts (1 tbsp.)

Preparation Method:

1. Whisk the water, sugar, salt, and vinegar in a small mixing container. Pop it into the microwave for one minute.
2. Mince and add the garlic and chili in the bowl and thoroughly mix it.
3. Chop and add the cucumber and sprinkle with the peanuts.

Thai Green Curry With Seafood

Servings Provided: 4

Time Required: 20 minutes

What is Needed:

- Unrefined peanut oil (2 tbsp.)
- Garlic cloves (3 minced)
- Green onions (5)
- Cilantro (3 tbsp. - divided)
- Thai green curry paste (6 tbsp.)
- Water (1.25 cups)
- Unsweetened coconut milk (13-14 oz. can)
- Kaffir lime leaves (2)
- Red jalapeño chile (1) OR Thai red chiles (2 small)
- Fish sauce - ex. nam pla or nuoc nam (1 tbsp.)
- Carrot (1 large/about 1 cup)
- Bok choy (4 cups)
- Uncooked medium shrimp (8 oz.)
- Bay scallops (8 oz.)
- Green or black mussels (1 lb.)
- Fresh basil (2 tbsp.)
- Steamed rice (2 cups)

Preparation Method:

1. Do the prep. Finely chop the dark green parts of the onion - separated from white and pale green parts. Mince the cilantro, garlic, basil, and onion.
2. Peel and thinly slice the carrot and bok choy. Peel and devein the shrimp. Scrub and debeard the mussels.
3. Prepare a large saucepan using the medium-temperature setting to warm the oil. Add the white and pale green parts of green onions,

44

garlic, and one tablespoon cilantro. Sauté them until tender (2 min.)

4. Add the curry paste, sautéing it for about one minute until it's fragrant.

5. Add coconut milk, water, lime leaves, chilies, and fish sauce. Once it's simmering, add the carrots and cover. Cook until the carrot is tender (5 min.).

6. Layer the bok choy, scallops, mussels, and shrimp in the pan. Cover and simmer them until the seafood and bok choy are cooked, and the mussels have opened. Discard any of the mussels that do not open after about five minutes.

7. Stir in two tablespoons of the cilantro, the dark green parts of green onions, and basil.

8. Portion rice among four shallow bowls. Ladle the curry over rice and serve.

Thai Grilled - Baked Fish

Servings Provided: 4-6

Time Required: 1 ¼ hours – varies

What is Needed:

- Frozen whole milkfish (1 @ 2.5-3.5 lb.)
- Sea salt (1.5 tbsp.)
- Oyster sauce (1/3 cup)
- Thai soy sauce (2 tbsp.)
- Sugar (.5 tsp.)
- Fresh lemongrass (1 piece - smashed)
- Chef's knife (For cleaning the fish)

Preparation Method:

1. Thaw the fish overnight in the fridge. Thoroughly rinse it using cool water.
2. Use the knife to make three gashes on each side of the fish. Cover the fish with sea salt and let it rest for ten minutes.
3. Set the oven temperature at 350° Fahrenheit.
4. Place the fish into a large sheet of heavy-duty foil.
5. Combine the soy sauce, oyster sauce, and sugar. Dump it over the top of the fish and wrap it, crimping the edges of the foil on top to seal it shut.
6. Bake them for 40 minutes. You can also choose to grill over hot coals

for 50 minutes.

7. Check for doneness by flaking the flesh with a fork. When it's fully cooked, the meat should be flaky and white, not opaque. Leave the fish in the foil - undisturbed - for ten minutes.

8. Serve in the foil, on top of a large platter of red or green leaf kale or lettuce. Serve with a serving of jasmine rice (see the recipe).

9. Note: How to Clean the Fish: Scale the fish outside near a water source. Hold the knife at a 30-degree angle, scraping the scales away from yourself. Make a 3-inch incision at the stomach area, near the bottom edge of the fish. Trash the guts and leave the head and fin intact (if desired). You can skip all of this and purchase the fish pre-cut in the market.

PART II

You will find many of the Chinese recipes will call for shoyu. Shoyu is the term broadly given to soy sauces that are made from fermented soybeans, wheat, salt, and water. In general, they are quite thin and clear and are excellent as an all-purpose cooking and table sauce. One of the best selling shoyu in the world is proclaimed to be Kikkoman Soy Sauce.

Chapter 1: Soup

Hot & Sour Soup

Servings Provided: 4

Time Required: 40 minutes

What is Needed:

- Chicken broth - low-sodium (1 quart)
- Dried tree ear fungus (.25 cup)
- Dried lily buds (12)
- Medium-dark soy sauce (2 tbsp. + more for seasoning)

- Distilled white vinegar (2 tbsp. + more for seasoning)
- Cornstarch (2 tbsp.)
- Kosher salt (.5 tsp.)
- Large eggs (2)
- Bamboo shoots (.5 cup - shredded)
- Cooked pork, ham, or chicken (.5 cup - shredded)
- Spiced thick - dry tofu - shredded (1 cup/3.5 oz.)
- White pepper - finely ground (1.5 tsp.)
- Sesame oil (1 tbsp.)
- To serve: Chopped cilantro & scallions

Preparation Method:

1. Dump the lily buds into boiling water to soak about ten minutes until they're softened. Discard the rough tips.

2. Prepare another container and add the tree fungus and boiling water to soak from 20 minutes to half an hour. Rinse, drain, and coarsely chop them.

3. Dump the broth into a large saucepan. Once boiling, add the soy sauce, salt, and vinegar.

4. Whisk three tablespoons of water with the cornstarch and mix it into the broth to simmer for three to four minutes to thicken.

5. Once it's at a rolling boil, whisk the eggs and a dash of salt, and work it into the soup in a circular fashion. Wait five seconds, stir and extinguish the heat.

6. Toss in the tofu, chicken, white pepper, bamboo shoots, ear fungus, and lily buds.

7. Simmer the soup using the medium temperature setting for about two minutes, adding in vinegar, soy sauce, and salt as desired.

8. Portion the soup and garnish it using the cilantro, scallions, and a spritz of sesame oil.

Wonton Soup

Servings Provided: 8

Time Required: 1 hour 15 minutes

What is Needed:

- Pork - fresh loin - whole (.5 lb.)
- Crustaceans - shrimp - mixed-species - raw (2 oz.)
- Brown sugar (1 tsp.)
- Burgundy wine (1 tbsp.)
- Shoyu - low-sodium soy sauce (1 tbsp.)
- Spring onions/scallions - tops & bulb (1 tsp.)
- Ginger root (1 tsp.)
- Wonton wrappers - includes egg roll wrappers (24 @ 3.5-inch square)
- Clear chicken broth - Swanson - CAM (3 cups)
- Scallions (includes tops and bulb- (1/8 cup)

Preparation Method:

1. Chop the green onion and add one teaspoon into a large mixing container and add the pork, shrimp, sugar, wine, shoyu sauce, and ginger. Thoroughly toss the mixture and let stand for 25 to 30 minutes.
2. Scoop the filling (1 tsp.) into the middle of each wonton skin.
3. Moisten the four edges of the wonton wrapper with a small amount

of water on your fingertips, and pull the top corner down to the bottom, folding the wrapper over the filling to create a triangle.

4. Seal it by pressing the edges firmly. Bring the left and right corners together above the filling and overlap the corner of the tips. Moisten with water and press together. Repeat the process until all wrappers are used.

5. Make the soup. Heat the chicken stock to a rolling boil. Add the wontons and cook for five minutes.

6. Top off the soup with chopped green onions and serve.

Chapter 2: Seafood

Honey Walnut Shrimp

Servings Provided: 4

Time Required: 30 minutes

What is Needed:

- Water (1 cup)
- English walnuts (.5 cup)
- Granulated sugar (2/3 cup)

- Egg white - raw (4)
- Rice flour - white (2/3 cup)
- Salad dressing/soybean oil with salt/mayonnaise (.25 cup)
- Jumbo shrimp - fresh & raw (1 lb./21-30)
- Honey - strained or extracted (2 tbsp.)
- Sweet condensed canned milk (1 tbsp.)
- Oil for frying (1 cup)

Preparation Method:

1. Whisk the water and sugar in a small saucepan. Once boiling, add the walnuts and boil them for two minutes. Dump them into a colander to drain. Arrange the nuts on a baking tray to thoroughly dry.
2. Whip the egg whites in a mixing container until they're foamy. Stir in the mochiko until it's a pasty consistency.
3. Warm the oil using the med-high temperature setting in a heavy deep skillet.
4. Dip the shrimp into the batter, and fry them until nicely browned (5 min.). Transfer them to a paper towel-lined platter using a slotted spoon to allow them to drain.
5. Whisk the honey, mayonnaise, and sweetened condensed milk. Fold in the shrimp and toss to coat with the sauce.
6. Garnish using the candied walnuts right before serving.

Steamed Fish

Servings Provided: 2

Time Required: 35 minutes

What is Needed:

- Raw finfish, snapper, mixed species (1 lb.)
- Salt (.5 tsp.)
- Black pepper (.5 tsp.)
- Ginger root - raw (1 tbsp.)
- Shoyu soy sauce (1 tbsp.)
- Sesame oil (2 tsp.)
- Shiitake mushrooms - raw AMM (2)
- Tomatoes (1 fresh)

- Peppers - raw red-hot chile (half of 1)
- Cilantro (2 sprigs - raw)

Preparation Method:

1. Prepare a steamer with a basket large enough for the snapper to lie flat. Pour in 1.5 inches of water and wait for it to boil.
2. Sprinkle the snapper with pepper and salt and pepper before placing it into the basket. Top the fish with ginger, and drizzle with sesame oil and soy sauce.
3. Place the tomatoes, mushrooms, and red chile pepper in the steamer basket.
4. Set a timer and steam the fish for 15 minutes, or until easily flaked with a fork. Garnish with cilantro and serve.

Stir-Fried Shrimp & Scallions

Servings Provided: 4

Time Required: 30 minutes

What is Needed:

- Jumbo shrimp (1.5 lb.)

- Garlic (3 cloves)
- Fresh ginger (1-inch section)
- Crushed red pepper (1.5 tsp.)
- Egg white (1 large)
- Cornstarch (2 tsp. - divided)
- Ketchup (.75 cup)
- Chicken broth - low-sodium (.5 cup)
- Black pepper & kosher salt (1.5 tsp. each)
- Sugar (1 tbsp.)
- Canola oil (.25 cup)
- Chopped cilantro (.5 cup)
- Scallions (3)

Preparation Method:

1. Thinly slice the scallions. Shell and devein the shrimp. Mince the garlic and ginger.
2. Toss the shrimp with the ginger, garlic, red pepper, one teaspoon of the cornstarch, and egg white until well-coated.
3. Whisk the broth with the ketchup, sugar, salt, and pepper with the rest of the cornstarch.
4. Warm a large skillet with the oil until it shimmers. Add the shrimp and stir-fry using the high-temperature setting until pink.
5. Add the ketchup mixture and simmer until the shrimp are heated (2 min.). Stir in the cilantro and scallions to serve.

Chapter 3: Poultry

Kung Pao Chicken - Keto-Friendly

Servings Provided: 4

Time Required: 40 minutes

What is Needed:

- Chicken breasts - boneless skinless (1 lb.)
  The Marinade:

- Chinese rice wine/dry sherry (2 tsp.)
- Soy sauce (2 tsp.)
- Cornstarch (2 tsp.)

  To Cook:

- Olive oil or sunflower oil (3 tbsp. divided)
- Dried red chilies (4-6)
- Green onions (4)
- Optional: Red finger chili (1)
- Asparagus (1 bunch)
- Sweet bell pepper (1)
- Garlic & ginger (4 tsp. each)
- Mini cucumbers (2)
- Roasted salted peanuts/cashews (.33 cup)
- Toasted sesame seeds (1 tsp.)

  The Sauce:

- Water (3 tbsp. - cold)
- Soy sauce & white vinegar (2 tbsp. each)
- Chinese wine/ sherry (1 tbsp.)
- Cornstarch (2 tsp.)
- Salt (.5 tsp.)
- Optional: Asian chili-garlic sauce (1 tbsp.)

Preparation Method:

1. Slice the chicken into one-inch chunks and combine it with the marinade fixings in the first group (soy sauce, rice wine, and cornstarch) stirring to combine. Marinate the mixture for 15 minutes.

2. Prep the veggies by cutting the asparagus into large pieces and mincing the garlic and onion. Chop the cucumber. Core and cube the bell pepper. Cut the green onions into one-inch pieces.

3. Prepare a large-sized cast-iron pan to warm one tablespoon of oil using the med-high temperature setting. Add the green onions, dried chilies, and finger chili.

4. Simmer the mixture until the green onions are slightly charred (1 min.). Transfer them to a baking sheet or large platter.

5. Heat the rest of the oil in the pan (1 tbsp.). Add asparagus and bell pepper and cook, stirring until it's slightly charred (2-3 min.).

6. Transfer to a baking tray and add the remainder of the oil (1.5 tsp.) to the pan. Working in two batches, stir-fry the chicken until browned (3-4 min. per batch), repeating with remaining oil.

7. Make the Sauce: Whisk the water, rice wine, soy sauce, and cornstarch until smooth. Return the chicken, vegetables, and chilies to the pan.

8. Sprinkle with salt, stir in the sauce, and cook until the liquid is bubbling and thickened (30 seconds to one minute). Stir in the cucumbers, peanuts, and chili-garlic sauce.

9. Serve it with a garnish of sesame seeds.

Orange Chicken

Servings Provided: 4

Time Required: 35 minutes

What is Needed:

The Chicken:

- Oil (as needed for frying)
- Boneless & skinless chicken breasts (4)
- Eggs (3 whisked)
- Cornstarch (.33 cup)
- Flour (.33 cup)

Orange Chicken Sauce:

- Orange juice (1 cup)
- Sugar (.5 cup)
- Rice/white vinegar (2 tbsp.)
- Tamari or soy sauce (2 tbsp.)
- Ginger (.25 tsp.)
- Garlic powder (.25 tsp.) or 2 garlic cloves (2 finely diced)
- Red chili flakes (.5 tsp.)
- Orange Zest (1 orange)
- Cornstarch (1 tbsp.)

The Garnish:

- Orange Zest
- Green Onions

Preparation Method:

1. Prepare the orange sauce. Mince the ginger and garlic. Pour the vinegar, orange juice, soy sauce, sugar, garlic, ginger, and red chili flakes into a saucepan. Sauté them for about three minutes.
2. Whisk one tablespoon of cornstarch with two tablespoons of water to form a paste. Add it to the orange sauce and whisk thoroughly. Continue cooking the sauce for about five minutes, until the mixture begins to thicken. After it's thickened, remove the pan from the burner and add the orange zest.
3. Prepare the chicken by cutting it into bite-sized chunks.
4. Dump the flour, a pinch of salt, and cornstarch in a pie plate or another shallow dish.

5. Whisk eggs in a shallow mixing container.

6. Dip the pieces of chicken into the egg mix and then flour mixture. Place them onto a platter.

7. Next, warm two to three inches of oil in a heavy-bottomed skillet (med-high temperature). Use an electric skillet or use a thermometer to check the heat until it reaches 350° Fahrenheit.

8. Working in batches, fry several chicken pieces at a time. Cook them for two to three minutes, often turning until golden brown, and place the chicken on a paper-towel-lined plate. Repeat the process until all the chicken is cooked.

9. Toss the chicken with the orange sauce. Reserve some of the sauce to serve over the rice. Serve it with a sprinkling of green onion and orange zest to your liking.

Chapter 4: Pork

Chinese Pork BBQ (Char Siu)

Servings Provided: 4

Time Required: 3 hours 40 minutes

What is Needed:

- Fresh pork tenderloin - lean cut (2 lb.)

- Soy sauce -shoyu - made from soy and wheat (.5 cup)

- Honey - strained/extracted (.33 cup)
- Ketchup (.33 cup)
- Brown sugar (.33 cup)
- Hoisin sauce - ready-to-serve (2 tbsp.)
- Rice wine (.25 cup)
- Red food coloring (.5 tsp.)
- Chinese Five-Spice Powder (1 tsp.)

Preparation Method:

1. Cut the pork "with the grain" into strips 1.5-2-inches long, and toss it into a large resealable zipper-type baggie.

2. Whisk the soy sauce, ketchup, honey, hoisin sauce, brown sugar, red food coloring, Chinese 5-spice, and rice wine in a saucepan using the med-low temperature setting. Simmer it until just combined and slightly warm (2-3 min.). Pour the marinade into the bag with the pork, pushing the air from the bag, and zip it closed. Toss the bag several times to cover all pork pieces.

3. Pop the pork in the fridge for two hours or overnight.

4. Warm the outdoor grill using the med-high temperature setting and lightly grease the grate.

5. Transfer the pork from marinade, shaking it to remove excess juices. Discard the marinade.

6. Grill the pork for 20 minutes. Place a container of water onto the grill and continue cooking, turning the pork until cooked thoroughly or about one hour. It's ready when the internal temp reaches 145° Fahrenheit.

Chinese Pork Dumplings

Servings Provided: 5/50 dumplings

Time Required: 1 hour 20 minutes

What is Needed:

- Soy sauce made from wheat & soy - shoyu (.5 cup)

- White rice vinegar CBT (1 tbsp.)

- Chinese chive - kucai - raw (1 tbsp.)

- Dried sesame seeds - whole (1 tbsp.)

- Sriracha sauce/Chili puree sauce w/Garlic CBT (1 tsp.)

- Freshly ground pork - raw (1 lb.)

- Garlic (3 cloves)

- Egg - whole (1)

- Kucai - Chinese chive - raw (2 tbsp.)

Preparation Method:

1. Combine ½ cup of the soy sauce, rice vinegar, sesame seeds, one tablespoon of chives, and the chile sauce in a small mixing container. Set it to the side for now.

2. Mix the pork, minced garlic, egg, two tablespoons of chives, soy sauce, sesame oil, and ginger in a large mixing container until thoroughly combined.

3. Lightly flour a workspace. Place a dumpling wrapper onto it and

spoon about one tablespoon of the filling in the center.

4. Wet the edge with a little water and crimp it together, forming small pleats to seal the dumpling. Repeat the process with the rest of the dumpling wrappers and filling.

5. Warm one to two tablespoons of oil in a large skillet using the med-high temperature setting. Arrange eight to ten dumplings in the pan and cook until browned (2 min. per side).

6. Pour in one cup of water, place a lid on the pot, and simmer until the pork is thoroughly cooked and the dumplings are tender (5 min.).

7. Continue the process until all dumplings are prepared. Serve with the soy sauce mixture for dipping.

Chop Suey

Servings Provided: 6

Time Required: 51 minutes

What is Needed:

- Fresh pork tenderloin (1 lb.)

- Wheat flour, all-purpose, white, enriched, bleached (.25 cup)

- Oil - soybean, salad or cooking (2 tbsp.)

- Bok choy - raw (2 cups)

- Celery - fresh (1 cup)

- Sweet red bell peppers (1 cup)

- Mushrooms (1 cup)

- Water chestnuts, Chinese, canned - solids & liquids (8 oz. can)

- Garlic (2 fresh cloves)

- Swanson Clear Chicken Broth CAM (.25 cup)

- Shoyu sauce (.25 cup)

- Cornstarch (1 tbsp.)

- Fleischmann's Cooking Sherry II (1 tbsp.)

- Ground ginger (.5 tsp.)

Preparation Method:

1. Use a sharp knife to discard the fat from the pork and slice it into one-inch pieces. Combine the flour and pork in a resealable bag, seal, and shake it thoroughly to cover.

2. Warm one tablespoon oil in a large skillet using the med-high heat setting. Add the trimmed pork and cook for three minutes or until browned. Transfer it to a container and keep it warm.

3. Pour the rest of the oil in the pan to heat. Add the celery, bok choy, mushrooms, red pepper, garlic, and water chestnuts. Stir-fry them for three minutes.

4. Thoroughly whisk the chicken broth, soy sauce, cornstarch, sherry, and ginger in a mixing container.

5. Combine the pork and broth mixture in a skillet, and cook for one minute or until thickened.

Easy Moo Shu Pork

Servings Provided: 6

Time Required: 1 hour 20 minutes

What is Needed:

- *Shoyu* - Soy sauce made from soy + wheat (2 tbsp.)
- Sesame oil (1 tbsp.)
- Garlic (1 tsp.)
- Fresh ginger root (1 tbsp.)
- Pork tenderloin (.75 lb.)
- Oil - soybean - salad or cooking (2 tbsp.)
- Chinese cabbage (pe-tsai) (2 cups)
- Carrots (1 raw)
- Salt (1 pinch)

Preparation Method:

1. Whisk the sesame oil, soy sauce, garlic and ginger in a bowl until the marinade is smooth. Dump it into a resealable plastic bag and add the pork. Cover it using the marinade, squeeze out any excess air, and seal the bag. Marinate in the fridge for a minimum of one hour to overnight.
2. Warm vegetable oil in a wok/large skillet using the medium temperature setting. Rinse and add the cabbage and diced carrot. Simmer the mixture for one to two minutes.
3. Push the cabbage mixture aside and add pork with marinade to the center of the skillet. Cook and stir until the pork is thoroughly cooked (3-4 min.).
4. Scoot the cabbage into the center of the skillet and continue to cook it for another minute or two. Adjust the flavor with a portion of pepper and salt to your liking.

Peking Pork Chops - Slow-Cooked

Servings Provided: 6

Time Required: 6 hours 15 minutes

What is Needed:

- Pork chops - top loin (6 boneless)
- Brown sugars (.25 cup)
- Ground ginger (1 tsp.)
- Shoyu soy sauce (.5 cup)

- Ketchup (.25 cup)
- Garlic (1 clove)
- Salt (as desired)

Preparation Method:

1. Use a sharp knife to remove the fat from the chops and toss them into the cooker.
2. Whisk the sugar, soy sauce, ginger, garlic, pepper, and salt. Dump it over the meat
3. Securely close the lid and set the timer for four to six hours.
4. Serve when it's tender with a dusting of salt and pepper as desired.

Chapter 5: Other Chinese Dishes

Crispy Tofu With Sweet & Sour Sauce

Servings Provided: 4

Time Required: 45 minutes

What is Needed:

The Sauce:

- Cornstarch (2 tsp.) + Water (2 tsp.)
- Garlic (2 minced cloves)
- Freshly grated ginger (.5 tsp.)
- Chili pepper flakes (.25 tsp.)
- Vegetable oil (2 tsp.)
- Water (.5 cup)
- Unseasoned rice vinegar (.33 cup)
- Agave nectar (.5 cup)
- Low-sodium soy sauce (2 tbsp.)
- Tomato paste (2 tbsp.)
- Sea salt (.25 tsp.)
  The Tofu & Batter:

- Medium/firm tofu (1 brick)
- For Frying: Vegetable oil (3 cups)

- Cornstarch (1 tbsp.)
- Brown rice flour (1 cup)
- Ground pepper (.25 tsp.)
- Sea salt (.5 tsp.)
- Garlic powder (.5 tsp.)
- Cold soda water (1 cup)

Preparation Method:

1. Drain the brick of tofu and chop it into bite-sized cubes. Continue to drain the cubes on a layer of paper towels to remove the excess water. Press it often while you prepare the sauce.
2. Mix water with the cornstarch in a cup and set it aside for now.
3. Warm two teaspoons of vegetable oil using the med-low temperature setting. Mince and add the ginger, garlic, and chili pepper flakes. Stir for 30 seconds to one minute until fragrant.
4. Whisk in the rest of the sauce ingredients using the medium setting until it's bubbly. Whisk in the cornstarch mixture.
5. Whisk the sauce often for 10-12 minutes until slightly thickened. Transfer the pan to a cool burner while you prepare the crispy tofu.
6. Warm three cups of oil in an electric skillet or pan to reach 375° Fahrenheit.
7. Mix the batter by combining the rice flour, cornstarch, sea salt, garlic powder, and ground pepper in a mixing container.
8. When the pan is hot, stir in the soda water to the flour mixture and mix well.
9. Use your hands to coat three to four cubes of tofu and gently place them into the oil. Fry them for 2-2.5 minutes.

10. Remove the tofu using a slotted spoon and place them onto a layer of paper towels to absorb the excess fat. Repeat the process with the rest of the tofu cubes.

11. Warm the sauce if needed. In two to three batches, you can coat the crispy tofu with sauce by adding a portion of the sauce to a large bowl and tossing the crispy tofu cubes until coated evenly. Serve to your liking with veggies or rice.

Shiitake & Scallion Lo Mein

Servings Provided: 8

Time Required: 40 minutes

What is Needed:

- Lo mein noodles (1 lb.)
- Snow peas (.25 lb.)
- Mirin (.25 cup)
- Soy sauce (.25 cup)
- Toasted sesame oil (2 tsp.)
- Canola oil (3 tbsp.)
- Shiitake mushrooms (1 lb.)
- Scallions (6)
- Fresh ginger (1 tbsp.)
- Water (2 tbsp.)
- Cilantro (2 tbsp.)

Preparation Method:

1. Slice the snow peas diagonally into halves. Remove the stems and thinly slice the caps of the mushrooms. Cut the scallions into one-inch lengths. Mince the ginger and chop the cilantro.
2. Prepare a large pot of boiling salted water. Cook the noodles until tender, adding in the snow peas for the last two minutes of the cooking cycle. Rinse and drain the noodles and snow peas in a

colander using cold water until cooled.

3. Whisk the soy sauce with the sesame oil and mirin.

4. Prep a deep skillet to warm two tablespoons of the canola oil until shimmering using the high-temperature setting. Add the shiitake and cook it, undisturbed, until browned (5 min.).

5. Add the rest of the canola oil, scallions, and ginger. Stir-fry until the scallions softened (3 min.).

6. Add the water into the pan and simmer using moderate heat, scraping up the browned bits from the bottom of the pan for about a minute or so.

7. Mix in the snow peas, noodles, and soy sauce mixture. Simmer while tossing the noodles until they are thoroughly heated (2 min.).

8. Sprinkle using the cilantro and transfer it onto banana leaf cones or bowls to serve.

PART III

Chapter 1 Introduction to the Low FODMAP diet

Do you suffer from abdominal cramping and discomfort? If you spend your days feeling constipated, bloated, and feel the uncontrollable urge to use the bathroom? If so, you may suffer from IBS.

With so many diets on the market, it can be hard to decide which one is best for you! In the following chapters, you will be learning everything you need to know about the FODMAP diet and how it can benefit your life.

Unfortunately, there are several theories behind why individuals suffer from IBS. For many, there is 70% of women who suffer from IBS due to their hormones triggering the symptoms. As for others, the reasons could be anything from a sensitive colon, an immune response to stressors, sensitive brain activity in detecting gut contractions, or even a neurotransmitter serotonin being produced in the gut. While the doctors are unable to pinpoint an exact reason for IBS, the good news is that they are certain that IBS will not cause other gastrointestinal diseases and it is not cancer!

The right question to ask in this moment, is what can I do about it? We are here to tell you that the low FODMAP diet is the way to go. In the chapters to follow, you will learn everything from what the diet is, who the diet is for, what FODMAP even stands for, and why this diet will work for you. We cover the benefits of the diet and include an easy start guide so you can get rid of that discomfort and bloat as soon as possible!

To start, it is important to understand what the FODMAP diet is, and why it is something you need to start. However, before we start, here are some tips for the beginners who are just starting or considering the low FODMAP diet.

Getting Started

Before we begin, it is important to get a diagnosis from your family doctor. Many people self-diagnose themselves with IBS and place themselves on the low FODMAP diet. This is something we do not recommend. If you have symptoms such as pain and bloating, you should see a professional to rule out any possible life-threatening diseases.

What is IBS?

As mentioned, be sure to see a professional to attain an official prognosis of IBS. If you suspect you do have Irritable Bowel Syndrome, realize that you are not alone. In fact, around 15% of the population in the United States suffer from IBS symptoms. While the symptoms do vary from person to person, the typical symptoms are as follow:

- Bloating
- Constipation
- Diarrhea
- Lower Abdominal Pain
- Lower Abdominal Discomfort

If you suffer from any of these, it is important to consult with your doctor the specific symptoms you have. This will be vital as there are three different types of irritable bowel syndrome. These include:

- IBS with Constipation

- Typically, IBS with constipation has symptoms including bloating, abnormally delayed bowel movements, stomach pains, and loose or lumpy stool.
- IBS with Diarrhea
  - Typically comes with symptoms including stomach pain, urgent need to use the bathroom, loose and watery stool
- IBS with alternating Diarrhea and Constipation
  - Due to the fact that there are several types of IBS, this makes it hard to determine a single drug treatment to help with the symptoms. As we mentioned earlier, you need to consult with a professional. Once you have done this and ruled out any other illnesses, it is time to take a look at your diet.

Who is the diet for?

Typically, the low FODMAP diet is meant for individuals suffering from IBS. The diet itself was created as one of the first food-based treatments to help relieve IBS symptoms. The good news is that up to 75% of patients who had IBS experienced symptom relief when they followed the low FODMAP diet. However, the diet is also helpful if you have any of the following:

- Digestive Disorder
- Gastroesophageal Reflux Disease (GERD)
- Crohn's Disease
- Celiac Disease
- Vegan Gut

- Bloating

Once you have determined that the low FODMAP diet could help your symptoms, it is now time to learn what FODMAP even stands for! This is going to be vital information to carry with you through your diet so you understand what you are eating and why your body is reacting the way it does!

We understand that there are many different types of diets out there. Some of you may be wondering, can I follow my current diet and still follow the low FODMAP diet? The answer varies depending on which you follow, and we will try to answer in a simple manner:

- Vegetarian/ Vegan
  - o Yes, this diet is more than possible to follow if you are vegan or vegetarian. With a few tweaks, you can find friendly options and still stick to your regular diet!
- Low-Salt
  - o If you follow a low-salt diet, this diet is doable for you. However, it will be vital that you learn how to read and follow food labels. Luckily for you, this is information also included in this book!
- Gluten-Free
  - o As you will be learning, the FODMAP diet does exclude wheat, which contains gluten. If you are gluten-free, this diet is easy to follow as you most likely will not be able to have it anyway!
- Kosher

o   If you have to eat kosher, you can still follow this diet. It will be up to you to find certain kosher foods, but after the elimination diet, you will be able to find the foods and still stick to your original diet.

History of the low FODMAP diet

Originally, the low FODMAP diet was developed by a team of scientists at the Monash University located in Australia. The original research was meant to investigate if the diet would be able to control IBS symptoms with food alone. The university established a food analysis program to study FODMAPs in both Australian as well as international foods.

In 2005, the first FODMAP ideas would be published as part of a research paper. In the paper, the hypothesis was that by reducing dietary intake of certain foods that were deemed indigestible, this could help reduce symptoms stimulated in an individual's gut's nervous system.

Over many years, research has shown that certain short-chain carbohydrates such as lactose, sorbitol, and fructose was the cause behind gastrointestinal discomfort. Once the basis of digestion was studied, the low FODMAP diet was created to help with these symptoms.

What does FODMAP stand for?

FODMAPs are typically found in foods that we consume every day. They are in onions, rye, barley, wheat, garlic, milk, fruits, vegetables, and more! As you can tell from this very small list (don't worry, we will cover more in the chapters to follow), they are in some of our more common foods! This is why it is so easy to feel bloated for some people, without understanding what is causing it! However, before we dive into how this diet works, you will need to understand the acronym FODMAP.

F- Fermentable

O- Oligosaccharies (short chain carbohydrates)

D- Disaccharides (lactose)

M- Monosaccharides (fructose)

A- and

P- Polyols (Sorbitol, xylitol, maltitol, and mannitol)

The reason you may be suffering from IBS or other digestive issues is due to the fact that most FODMAPs have a hard time absorbing into your small intestine. As a result, these FODMAPs are fermented by the bacteria in your small and large intestine in which results in bloating and irregular bowel movements.

While the FODMAPs cause the digestive discomfort, it is important to understand that it is not the cause of the intestinal inflammation itself. In fact, the FODMAPs produce alterations of intestinal flora that help you maintain a healthy colon. This does not change that the symptoms are still uncomfortable.

What may be causing your IBS symptoms could be a fructose malabsorption or a lactose intolerance. As you will be learning in later chapters, as you begin the low FODMAP diet, there will be an elimination phase where you learn what exactly is causing your symptoms and discomfort.

The source of the FODMAP will vary depending on different dietary groups. In more common circumstances they are compromised as the following:

- Oligosaccharies: Fructans and Galacto-oligosaccharies
- Disaccharies- Lactose
- Monosaccharies- Fructose
- Polyols- Xylitol, Mannitol, Sorbitol

Sources of Fructans

In later chapters, we will be going more in-depth on the foods you can and cannot eat. To cover the basics, you should understand where these specific irritants come from. To start, we will go over the source of fructans. These can be found in very popular ingredients including; rye, garlic, onion, wheat, beetroot, Brussel sprouts, and certain prebiotics.

Sources of Galactans

As for galactans, these are primarily found in beans and pulses. It can also be found in certain tofu and tempeh, but this does not mean that vegans and vegetarians cannot follow the low FODMAP diet. It simply means that you will need to find other sources of proteins if you want to follow a plant-based diet. We will be going over this more in the chapters to follow.

Sources of Polyols

Polyols are typically found in stone fruits. These include avocados, apples, blackberries, watermelon, and more. They are also found naturally in certain vegetables and bulk sweeteners.

While this diet may seem to be lacking many of your favorite foods, don't you worry! Due to the wide variety of IBS symptoms, it is unclear which foods trigger certain individuals. This is why the elimination trial will be important before you start the diet. Please remember that everyone is different. While some people see immediate results when they begin the diet, for others, it will take some time.

Effectiveness and Risks of the low FODMAP diet

It is important to understand that the low FODMAP diet is meant for short-term symptom relief. However, long-term diet can have a negative effect on your body. Unfortunately, it can be detrimental to your guy metabolome and microbiota. It is to be taken very seriously that this diet is meant for short periods of time and only under the advice of a professional.

Please understand that if you choose to follow the low FODMAP diet without any medical advice, it is possible the diet could lead to some serious health risks. Some of these risks are as followed:

- Nutritional Deficiencies
- Increased Risk of Cancer
- Death

When you start the low FODMAP diet, it is possible the diet itself could mask any serious disease that present themselves of digestive symptoms. These could include celiac disease, colon cancer, or inflammatory bowel disease. This is why it is so crucial to seek professional help before starting the diet on your own.

Now that you have learned the basics of the low FODMAP diet, it is time to learn all about the benefits that come with the diet change. Obviously, the main change will be to help lower any digestive troubles you may behaving. By removing the potential triggers in which are causing your digestive issues, this will help pinpoint which food intolerances you have.

While this diet may seem to take a lot of time and effort, think of the time you are wasting by being in discomfort all of the time and using the bathroom! With a few minor adjustments and tests, you will be able to find the source of your problem and hopefully never feel this way again! Now, onto learn all of the other incredible benefits the low FODMAP diet can bring to you!

Chapter 2: Benefits of the Low FODMAP diet

According to research, the low FODMAP diet is effective for around 75% of patients who suffer from IBS. In most cases, the patients are able reduce any major symptoms they are experiencing and in hand, improve their quality of life.

In the same research, scientist found evidence that the diet can also be beneficial for people who suffer from other functional gastrointestinal disorders such as Chron's disease, ulcerative colitis, and inflammatory bowel disease. All you need to do to benefit from this diet is to figure out what is causing the digestive disturbances and symptoms. Below, you will find some of the other benefits the low FODMAP diet has to offer:

A. IBS Symptom Reduction

By following the low FODMAP diet, individuals can reduce most symptoms involved with IBS including stomach pain, bloating, and gas. It is important to follow the diet and remove any irritants as they ferment inside of your intestines. By selecting foods that don't trigger your symptoms, you can avoid them altogether!

B. Chron's Disease Reduced Discomfort

By following the low FODMAP diet, individuals were able to change the quality and number of prebiotics. By controlling the foods you consume and avoiding the ones that trigger your system, you could reduce the discomfort you feel from the trigger foods.

C. Increased Energy

Some individuals feel tired no matter how much they eat through the day. It is believed that a low FODMAP diet can help reduce fatigue. This could be due to the fact that the body is no longer wasting energy on digesting foods that don't agree with your system. This is especially true for sweeteners that you could be using on a daily basis. As you will be learning later, some of the best sweeteners can be found in fruit!

D. Reduced Constipation and Diarrhea

When you follow the low FODMAP diet, you will begin to eliminate foods that are causing your symptoms in the first place. When you do this, your body will find a balance, and you may find that your bloating will decrease, the gas will decrease, and your stools will return to normal. It is a win-win situation! All you will need to do is figure out your triggers (which we cover in the third chapter) and follow the diet!

On top of these incredible benefits, there is also beliefs that the low FODMAP diet can benefit psychological health. Often times, the disturbances of IBS can cause stress to certain individuals, eventually leading to anxiety and depression. When you remove the trigger causing the symptoms by diet, you will be able to improve the quality of your life.

As you will be learning in the chapters to follow, the low FODMAP diet includes an elimination diet in order to get started. As you introduce foods, you may find that you have a lactose or gluten intolerance. While this may seem like a huge change, there are some incredible benefits to changing your diet.

Benefits of a Grain Free Diet

As you will be learning later in the book, there are many types of grains that has been found to cause inflammation. Unfortunately, this is a very common culprit for the digestive disorders you may be experiencing. This is especially true if you are very sensitive to gluten. The good news is that if you are looking to lose weight on top of feeling better, cutting out these grains will be the best thing to ever happen to you. Reined grains are high in carbohydrates and calories; they offer little to no nutrition and contribute to discomfort in your stomach. Before you make the switch to going grain free, consider some of the amazing benefits as follow:

A. Digestion Benefits

Gluten is a type of protein that can be found in wheat products. If you do the elimination diet and find that you are gluten sensitive, cutting it out makes the most sense. When you cut it out of your diet, this can help relieve issues such as nausea, bloating, diarrhea and constipation.

B. Reduced Inflammation

When you experience acute inflammation, this normally means that your immune system is fighting off foreign invaders. Unfortunately, if you sustain these levels for a long period of time, this is what causes chronic disease. By cutting gluten out, you can reduce the amount of inflammation in your body.

C. Balanced Microbiome

By following a grain-free diet, you will be able to balance the microbiome in your gut. When you do this, it helps support the beneficial bacteria in your body, helping improve your digestion, boost your immunity, and helps keep blood sugar under control.

D. Weight Loss

As mentioned earlier, most grains offer little to no nutrition. When you cut these extra calories out of your diet, it will help you lose weight. Instead of grains, try eating nutrient-dense foods like vegetables or legumes. Of course, you will figure out exactly what you can eat on the low FODMAP diet after going through the elimination portion.

Another common irritant when individuals suffer from IBS and other digestive orders can be lactose! You may be thinking to yourself; I could never give up my yogurt or ice cream. The good news is that in the current market, some incredible alternative choices can fit in the low FODMAP diet. In case you need some further benefits to help convince you, here are just a few:

E. Healthier Digestion

You may not know, but around 70% of the population has a degrees of intolerance to lactose. When we first begin to wean off of our mother's milk, we begin to use lactase. Lactase is an enzyme that helps digest lactose found in milk. As we age, we begin to lose the ability to digest lactose and is one of the biggest known triggers for IBS. By taking dairy out of your diet, you save yourself the troubles all together!

F. Decreased Bloating

91

Bloating occurs when we have issues with digestion. Some dairy products can cause excessive gas in the intestines, which is what causes the bloating in the first place. Some bodies are unable to break down the carbohydrates and sugar fully which in turn, creates an imbalance of gut bacteria.

G. Clearer Skin

If you suffer from acne, dairy could be the culprit! According to studies posted in Clinics in Dermatology, it was found that dairy products such as milk contain growth hormones that stimulate acne. By following the low FODMAP diet and cutting dairy from your diet, you could naturally treat acne!

H. Reduce Risk of Cancer

A 2001 study at Harvard School of Public Health found that there was a connection between high calcium intake and increased risk of prostate cancer. It is thought that the hormones in the milk contain contaminants such as pesticides that have been linked to cancer cell growth. These contaminants are mostly found in dairy products, giving you another reason to cut them from your diet altogether!

I. Decreased Oxidative Stress

It is believed that a high milk intake is typically associated with higher mortality rates in both men and women. This may be due to the D-galatose found in milk which helps influence oxidative stress and inflammation in the body. Unfortunately, this undesirable effect caused by milk can cause chronic expose and damage health. On top of inflammation, it can also shorten life spans, cause neurodegeneration and also decrease one's immune system.

As you can tell, there are so many incredible benefits of switching over to the low FODMAP diet. Whether you are looking to get rid of bloating, lose some weight, or stop constipation and/or diarrhea, the low FODMAP diet has got you covered. It is all a matter of figuring out what your trigger is in the first place.

Obviously, we could go on and on about the incredible benefits of the diet, but then we would never get to the diet itself! Now that you are aware of just some of the benefits, it is time to get you started! In the chapter to follow, you will be learning how to get started on the low FODMAP diet. You will the steps to get started on the diet itself and how to diet whether you are vegan, vegetarian, diabetics, or doing this for a child who suffers from IBS. When you are ready, we can dive in!

Chapter 3: Starting the Low FODMAP diet

Now that you are aware of the low FODMAP diet and some of its benefits, it is time to learn how you can get started on the diet yourself! While a diet and lifestyle change can seem daunting, it will be important to believe in yourself and remember why you started it in the first place. In the chapter to follow, we will be providing you with all of the information you need. From diagnosis of IBS, to starting the diet, and even how to practice if you are vegan, vegetarian, or diabetic. This diet can be universal; it is all about finding what works best for you. First, it is time to understand the diagnosis of IBS.

Getting Diagnosed with IBS and FODMAP Tests

If you are in the process of being diagnosed with a chronic medical condition, this could be a challenging time for you. It is important that you understand the symptoms the doctors are looking for, and which medical tests you will be taking in order to be officially diagnosed with irritable bowel syndrome. IBS can be diagnosed with a combination of Rome IV criteria so the doctors will be able to rule out any other gastrointestinal disorders.

Rome IV is a set of criteria that doctors have found that most IBS patients have in common. This criteria is 98% accurate when the doctors are identifying their patients with IBS. These criteria are as followed:

1. Recurrent abdominal pain at least one day per week in the last three months are have the following:

   A. Related to defecation

94

    B. A change in stool frequency

    C. Change in form of stool

2. Criteria from above is fulfilled with symptoms for at least six months before official diagnosis

Other symptoms often associated with IBS include bloating, abdominal pain, and a change in bowel habit. Your doctor will take in the evidence and match with the Rome IV criteria and will move onto discussing any red flag symptoms that may be occurring.

Before being diagnosed with IBS, it will be important that your doctor rules out any other medical conditions that could be presented with the same symptoms; this is where the red flag symptoms come into play. These flags include:

- Inflammatory Markers
- Rectal Masses
- Abdominal Masses
- Anemia
- Nocturnal Symptoms such as waking up from sleep to defecate
- Family History of coeliac disease, inflammatory bowel disease, and ovarian cancer
- Rectal Bleeding
- Unintentional Weight Loss

On top of these symptoms, your doctor will also ask for several other symptoms in order to diagnose you with IBS. Firstly, the discomfort and pain in your abdomen will need to be related to altered bowel frequency as well as a change in your stool form. You will also need to have at least two of the following symptoms:

1. Feeling incomplete emptying when using the bathroom
2. Passage of mucus when using the bathroom
3. Straining, Urgency, and Altered Stool passage
4. Abdominal Bloating
5. Lethargy
6. Backaches
7. Bladder Symptoms
8. Nausea

Medical Tests for IBS

Once your doctor figures out your symptoms, rules out any other serious medical conditions and believes it is appropriate to run tests for IBS can you expect one of the following tests:

- Antibody testing for coeliac disease
- C-reactive protein
- Erythrocyte sedimentation rate
- Full blood count

While these tests are typical, it may be different if you present any of the red flag symptoms from above. If you do have a red flag symptom, there will be additional tests to rule out any more serious issues. These tests are as follow:

- Hydrogen Breath Test (meant for lactose intolerance)
- Fecal Occult Blood

- Fecal Ova and Parasite Test
- Thyroid Function Test
- Rigid/ Flexible Sigmoidoscopy
- Ultrasound

In the case that your doctor feels your symptoms may not be linked to IBS, they will most likely refer you to a gastroenterologist. This is a physician who is an expert in managing diseases found in the liver and gastrointestinal tract. However, this is worst case scenario. For now, we will focus on following the diet if you are diagnosed with IBS.

Breaking down FODMAPS

In general, FODMAPs naturally occur in popular foods such as vegetables, fruits, grains, cereals, dairy products, and legumes. Unfortunately for those who suffer from IBS, these FODMAPs are absorbed poorly in our small intestines and can affect our bowels as a symptom. FODMAPs are short-chain carbohydrates found in these foods, but this does not mean that the diet itself is sugar-free. When we consume FODMAPs, they are fermented by gut bacteria in the large intestine in which triggers the unpleasant GI symptoms that you may be experiencing. Before we move onto the elimination stage, it is important to understand just what this acronym means.

Fermentable

Fermenting is the process where our gut bacteria attempt so break down FODMAPs. As you are already aware, these FODMAPs are indigestible carbohydrates and in turn, produce gas.

Oligo-Saccharies

This group of the FODMAP are broken down into two subgroups including fructans and galactans. Fructans also known as fructo-oligosaccharies or FOS are most commonly found in foods such as dried fruit, barley rye, wheat, garlic, onion. Galactans or galacto-oligosaccharies or GOS are found in pulses, legumes, cashews, pistachios, and silken tofu. If you feel yourself panicking over remembering these foods, don't worry. In the chapter to follow, we will cover exactly what you can and cannot eat while on the low FODMAP diet.

Di-Saccharies

As mentioned in the chapter from before, lactose could be a potential trigger in your diet. These can be found in any product that comes from goat, sheep, or cow's milk. Lactose itself contains two sugars united that require an enzyme known as lactase before our bodies are even able to absorb it. When your gut lacks these enzymes, this is when you can trigger symptoms of IBS.

Mono-Saccharides

This is a fructose that is found when a person has an excess amount of glucose in their diet. Our bodies need an equal amount of glucose in our system to stop any malabsorption. This means that while some of us can consume a certain amount of glucose, it is important to avoid foods that contain an excess amount. Some examples of these excessive foods include asparagus, honey, apples, and pears.

And Polyols

Polyols are also known as sugar alcohols. These can be found in a wide range of vegetables and fruits including sweet potatoes, mushrooms, pears, and apples. These sugar alcohols are also found artificial sweeteners in chewing gum, diabetic candy, and even protein powder. These polyols can only be partially absorbed into our small intestines. The rest continue into the large intestine, begin to ferment, and cause discomfort and bloating for some people.

As you begin to consider the low FODMAP diet, it is important to understand that one size does not fit all. This diet will change depending on your intolerance to certain foods. On the FODMAP diet, you will be following three different phases including: The Elimination Phase, The Reintroduction Phase, and the Maintenance Phase. We will go further into detail of each phase, so you have a full understanding before beginning.

The Elimination Phase

The Elimination phase is also known as the restriction phase. While this may seem intimidating, realize that this phase is only meant to last two to six weeks. This phase should only last long enough for you to gain control over your symptoms. Once this happens, you will move onto the reintroduction phase with the help of a professional. It is important that this stage is short as it can have long-term effects on your gut health.

To begin, you will want to create a personal list of foods you feel makes your IBS worse. If you are unsure which foods could be causing your symptoms, you will want to check out the next chapter to see an extended list of foods you should be avoiding. Some popular starters include chocolate, coffee, nuts, and certain fibers.

Once you have made your list, you will begin to eliminate these foods one at a time from your diet. It will take a couple of weeks before you notice any improvements. It does take some time for these foods to get through your system. However, if you do not notice any improvement, you will want to reintroduce these foods into your diet and try the next item on your list. Eventually, you will have a complete list of foods that trigger your IBS symptoms. Other popular foods to eliminate during this phase include: soy, gluten, and dairy products.

An important tool during this phase will be a food diary. This way, you will be able to keep track of which foods you are eating during the day, and any symptoms that may present themselves after they have been consumed. In general, the longer this phase is, the more likely you are to find that is triggering your IBS symptoms. It is important to remember that once eliminated, you will need to reintroduce foods slowly in the next phase.

The Reintroduction Phase

Once you have gone through your elimination period, you will be reintroducing these targeted foods back into your diet. While following the low FODMAP diet, you will need to introduce these foods back into your lifestyle one at a time.

As a tip for the reintroduction phase, we suggest starting on a Monday. This way, you will be able to consume a small portion of the food, wait a few days, and see if you experience any symptoms. On day three, you can eat a larger portion and wait another couple days for any onset symptoms. Be sure to keep track of how you are feeling in your food diary so you can present it to a professional if need be. If you experience symptoms, this is a possible food trigger. If there is no symptom, you can assume that this certain food group is a good match for your diet.

After a while, you will have a complete list of foods that you need to assess, and you will start the elimination phase over again to double check. Once you eliminate and reintroduce, you will be able to create a diet you can stick with and eliminate any symptoms of IBS you may experience.

Maintenance Phase

While this may take time, the elimination and reintroduction phase are going to be very important while following the low FODMAP diet. These are going to be your tools in identifying foods that trigger your IBS symptom. The long-term goal is to create a wide variety of foods you can consume on a daily basis to ensure you are intaking all of your essential nutrients while eliminating the ones that make you feel lousy.

As you go through these phases, it is vital you listen to your body. Only you will be able to tell if you have a tolerance to certain foods. Remember that portion sizes will be important during these phases as well. While you may not react to small portions, larger portions may trigger the symptoms which you will want to avoid. The more tests you do, the more foods you will be able to add or subtract from your diet. While this may take some extra work, it will be worth it when you decrease your bloating and discomfort from IBS.

You may be wondering if you can follow this diet even if you have a certain lifestyle. Typically, the answer is yes! The only factor being that you could have a very limited number of foods allowed on your diet with any other limitations. Below, we will cover some of the more common diets and how you can also follow the low FODMAP diet as well!

Low FODMAP Diet with Vegan/Vegetarian Diet

If you follow a vegan or vegetarian diet, you may want to consider working with a dietary professional. Due to the fact you consume a diet that is different from most of the population, it can be more difficult to access foods that can work well with both diets. By working with a professional, they can ensure you still follow your diets without missing any essential nutrients your body needs.

While on the low FODMAP diet, it is important you keep re-testing foods. Remember that the elimination phase is meant to be short term. As you reintroduce old foods, you will be able to process if you can tolerate them or not. While you do this, you can find some staple foods, even if they happen to be high in FODMAPs.

If you follow a vegan or vegetarian diet, it will be vital that you pay special attention to your protein intake. As you will be learning later in the book, the low FODMAP diet includes a limitation of many legumes which may be a main source of protein for you right now. Instead of legumes, you can consider soy products or simply a smaller portion of legumes. Along with these switches, there are also milk substitutes to help with your protein intake. There is almond milk and other soy products to help out. Certain nuts and seeds also have varying levels of proteins for you to consider.

Low FODMAP Diet and Diabetes

If you have diabetes, you are most likely aware that there is no specific diabetic diet. In general, most people with diabetes follow a suggested balanced and healthy diet. If you wish to follow the low FODMAP diet while having diabetes, there are a few key rules you can follow to ensure you do not cause further harm to your health.

1. Planning

While on the low FODMAP diet, planning regular meals will be key. By doing this, you will be able to make sure that your blood glucose levels are always stable. By planning in advance, you can be successful in managing your diabetes while still following the low FODMAP diet. This stands especially true if you struggle finding healthy foods when you are away from your house. By being prepared, you will always have healthy options and can stay away from temptations. One good idea is to prep snacks for in-between your meals. These can be rice cakes, popcorn, or a simple fruit that is allowed in your low FODMAP diet.

2. Focus on Low FODMAP Carbohydrates

If you wish to follow the low FODMAP diet and eat healthy while having diabetes, eating starchy carbohydrates will be important for you. Some suitable options for you include wheat free bread, oats, potatoes, and rice. Before you include these, be sure to eliminate them from your diet to assure they are not triggers. Essentially, you will want to avoid any large portions of carbohydrates so you will be able to avoid any spikes in your blood glucose. You can do this by choosing slow-release carbohydrates like sweet potatoes or oats. On top of these carbs, you will also want to include allowed vegetables.

3. High Sugar Foods

As a person with diabetes, you already know that sugary foods cause your blood glucose levels to rise. On the low FODMAP diet, there is a low risk of consuming sugary foods such as soft drinks and cake, but they should still be avoided.

4. Low FODMAP Fruit

While fruits are a source of sugar, it will be important that you include a few portions of fruit per day. On the low FODMAP diet, there are plenty of options such as grapes, strawberries, bananas, and even oranges. You will want to pay special attention to your portion sizes as bigger portions, means higher amounts of fructose. You will also want to limit your portions of dried fruit, smoothies and fruit juices as they are typically pretty concentrated sources of fructose.

Low FODMAP Diet for Children

At this point, there has been very little research on the low FODMAP diet for children. Studies have shown that there are no real negative side effects for individuals who follow the low FODMAP diet for short period of time. However, if this diet were to carry on for longer than suggested, it could possibly have a negative effect on the gut flora balance in a child. If you are considering this diet for your child, there are several factors you will want to take into consideration.

First, your child will need to be seen by the pediatrician to confirm that your child has irritable bowel syndrome. Once it is diagnosed, the doctor will need to approve the diet and be carefully supervised to assure the safety of your child. Only after you follow these steps should you continue onto the elimination stage of the low FODMAP diet. For success on the diet, you can follow some of the following tips:

1.   Inform Other Adults

Just like with any other diet restrictions, you will want to inform key adults of your child's restrictions. Whether it is a friend, a child care provider, or a teacher, this will be vital for the success of the diet. When these adults are in know of your child's diet, they will be able to address any stomach issues they may be having.

2.   Involve Your Child

If your child is old enough, try to explain the diet to them in simple terms. You will want to explain that they are feeling sick due to the food they are eating. Be sure to include them and ask for their input in the food substitutions and menu. By making your child feel they are a part of the process, this may help your child comply with the new food rules.

3.   Pack and Plan

Many parents fear diets for their children as they are always on the go! Luckily, the FODMAP diet is pretty easy to follow when you plan ahead. When you are at home, you most likely stock the fridge with low FODMAP foods. By planning ahead and packing your own snacks and lunches, you can assure your child will stick to the diet, so they do not make themselves sick.

4.   Forget the Small Stuff

Your kid is going to be a kid. If your child eats a restricted food every once in a while, it isn't going to ruin their diet altogether. Children typically do not have the self-discipline that adults have. They will most likely be tempted by restricted foods when at school or with their friends. You need to remember that while you want to stick to the diet most of the time, you can still allow your child some freedom when it comes down to what they are eating.

Exercise on the Low FODMAP Diet

While your diet may be causing your IBS symptoms, research has found that exercise can also help decrease any symptoms you may be suffering from. There are a few reasons why including regular moderate exercise will be important in the success of your diet.

First off, regular exercise can help reduce stress in your body. Typically, IBS tends to stress people out. When this happens, the nerves in your colon become tenser and can create abdominal pain. When your colon is tenser, this can slow down your bowel movements all together and cause constipation. A simple exercise such as cycling or walking can help release endorphins into your system and help release the tension in your colon. The more relaxed you are, the more flexible you will become.

Along with decreased stressed will come an increase of oxygen in your body. There are plenty of wonderful exercises such as tai chi and yoga that creates a breathing routine. When you take in these abdominal breaths, this helps increase the amount of oxygen in your body. As you increase oxygen, this will also help release any tension you are holding in your colon.

Finally, exercise can also increase your blood flow. As you begin to sweat, your body will be getting rid of toxins that could be creating discomfort in your colon. The more you sweat, the healthier you will be. Plus, the movement could help promote healthier bowel movements by moving blood to any problematic areas you may have.

As you consider exercise with your diet, remember that it will be vital to fuel your body before and after exercise. You will want to fuel about one to two hours before you work out. As long as it is included in your low FODMAP diet consider a banana with peanut butter or even oatmeal with some strawberries. The exercise can be any moderate activity of your choice from dancing, to running, to cycling, or even a little bit of strength training. Choose an exercise that makes you happy and one that you will stick with.

Reasons the Diet May Not Be Working

Speaking of sticking to a diet, some of you may follow these instructions and still suffer from IBS symptoms. If this still happens, you will want to take a look at your stress levels and the diet itself. While of course there is going to be a learning curve of the low FODMAP diet, allow yourself several weeks to change your food habits. Feel free to check back to the resources of this book to assure you are eating the foods allowed on the low FODMAP diet. If you still have no idea why you are experiencing the symptoms still, perhaps it is one of the following reasons that the diet isn't working:

1.   Lack of Fiber

Fiber plays a very important role in keeping your stool regular. Often times, the low FODMAP diet will remove high fiber foods, which means you will need to pay special attention to your fiber intake. If you find yourself struggling, try speaking to a professional to find other options while on the low FODMAP diet. It will also be important that you drink plenty of water to move fiber through your system.

2. Too Much Fruit

While there are plenty of fruits on the low FODMAP diet, it is possible you are eating too much of it in one sitting. Typically, you will want to stick to only one serving at a time. If you want more fruit later in the day, try waiting two to three hours after the first one is consumed. As you practice this diet more, you will be able to tell your tolerance levels with the fruits so you can reduce that time in between servings.

3. Hidden FODMAPs

Often times, you could be consuming ingredients that are high in FODMAPs and have no idea. Typically, they are found in highly processed foods to help their taste and texture. FODMAPs are also found in some medications such as cough drops and cough syrup. Unfortunately, while they can help a cold, they are often high in sugar alcohols which can trigger your IBS symptoms. It will be important to read labels, which is included in the chapter to follow.

4. Portion Control

It is very easy to sit down and eat more than a portion. While on the low FODMAP diet, allowed foods can become high FODMAPs when you exceed their allowed portion size. As an example, you may want to enjoy some rice cakes as a treat. A recommended serving size is only two rice cakes. If you eat double the allowed portion, this is when you may experience symptoms of IBS. Again, this is where reading labels carefully will come in handy while on the low FODMAP diet.

5. Stress

Stress is going to be a huge factor on the low FODMAP diet. If you are carefully following your diet, check your lifestyle. Stress itself can cause IBS symptoms so you may want to consider stress management skills along with a diet. You can try therapy or yoga. At the end of the day, your success is in your own hands.

If you continue to have IBS symptoms after following the diet and dealing with the issues from above, you may want to seek medical advice again. It is possible you have further intolerances that have not been explored yet. Also, the FODMAP diet will not work for everyone. If you have tried and failed, ask your doctor or dietician what the next step for you could be. For now, we will begin to cover the foods you can and cannot eat while on the low FODMAP diet.

Chapter 4: Low FODMAP diet foods

In the chapter to follow, you will find a list of both low and high FODMAP foods. As for the elimination phase, you will want to try to eliminate all of the high FODMAP foods. Once you are in the reintroduction stage, you will be able to introduce these foods back in order to see what is triggering your IBS symptoms.

As you choose your foods for your low FODMAP diet, remember that reading the ingredient list on a package is going to be vital for your diet success. Below, we will cover some of the basics of reading a food label. Too often, companies are able to hide food ingredients and could be triggering your symptoms without understanding why.

When you choose your foods, portion control will also be vital. When it comes to fruit, try your best to portion out one piece every few hours. As for processed foods, you will want to avoid them all together. If you ever have any doubts on low and high FODMAP foods, you can always revisit this chapter!

Reading and Understanding Nutrition Fact Label
If you are looking to eliminate certain foods from your diet, you will be surprised to learn that they can sneak into dishes without even realizing they are there. In order to stick with your diet, learning how to read and understand a nutrition fact label is going to be crucial for your diet.

A. Serving Size

When you first look at a label, you will want to check out the serving size along with the number of servings in any given package. These serving sizes are typically standardized so you can compare them to other similar foods. Remember that for some people, they can have smaller portions of FODMAP foods, but bigger portions could trigger IBS symptoms. When you are aware of a true serving size, this will make sticking to your diet a bit easier.

B. Calories

If you are on the low FODMAP diet to lose weight, this could be helpful for you. The calories in each package provide a measurement of how much energy comes in a serving of the food. The more calories you consume, the more you will gain weight. By being mindful of the calories in a portion, you will be able to manage your weight in a healthy manner.

C. Nutrients

When you look at a label, the first ones listed are typically the ones that Americans eat a good amount of. These can include Total Fat, Saturated Fat, Trans Fat, Cholesterol, and Sodium. While this isn't the main focus of the low FODMAP diet, it is something you should be mindful of for your general health.

D. Ingredients List

Finally, you will want to pay special attention to the ingredients list included on the package. If you are intolerant to certain ingredients, you will want to keep a food journal of these foods, so you always have them at hand to compare to a label. When looking at the ingredients list, they will be listed in order of weight from most to least. Eventually, you will know exactly what you can't eat and be able to compare easily in the store. As a beginner, remember to read the label of everything you put in your shopping cart.

When you understand the basics of reading a label, it is time to move onto learning the high and low FODMAP food list. We will begin with the high FODMAP foods. With this list, you will either want to avoid the foods altogether, or reduce them drastically. Of course, everyone's tolerances will be different but to help reduce any symptoms of IBS, you should reduce the following foods to enhance your health.

High FODMAP Foods (Avoid/ Reduce)

Fruits (High Fructose)

- Apples
- Avocado
- Apricots
- Blackcurrants
- Blackberries
- Boysenberry
- Currants
- Cherries
- Dates
- Figs
- Feijoa
- Guava
- Grapefruit
- Goji Berries
- Lychee
- Mango
- Nectarines
- Prunes
- Pomegranate
- Plums
- Pineapple
- Persimmon
- Pears
- Peaches
- Raisins
- Sultana
- Tamarillo
- Watermelon

Vegetables/ Legumes

- Asparagus
- Artichoke
- Butter Beans
- Broad Beans
- Black Eyed Peas
- Beetroot
- Bananas
- Baked Beans
- Choko
- Celery

- Cauliflower
- Cassava
- Fermented Cabbage
- Garlic
- Kidney Beans
- Leek
- Lima Beans
- Mushrooms

- Mixed Vegetables
- Pickled Vegetables
- Peas
- Red Kidney Beans
- Soy Beans
- Shallots
- Scallions
- Split Peas

Cereals and Grains

- Almond Meal
- Amaranth Flour
- Breadcrumbs
- Bread
- Biscuits
- Barley
- Bran Cereals
- Crumpets
- Croissants
- Cakes
- Cashews
- Cereal Bars
- Couscous
- Egg Noodles
- Freekeh

- Gnocchi
- Muesli Cereal
- Muffins
- Pastries
- Pasta
- Pistachios
- Udon Noodles
- Wheat Bran
- Wheat Cereals
- Wheat Flour
- Wheat Germ
- Wheat Noodles
- Wheat Rolls
- Spelt Flour

Sweeteners/ Condiments

- Agave
- Fruit Bar
- Fructose
- Hummus
- Honey
- High Fructose Corn Syrup
- Jam

-
-
- Kei...
- Sugar-Free Swee... (Inulin, Isomalt, Lactitol, Maltitol, Mannitol, Sorbitol, Xylitol)
- Tahini Paste

Drinks

- Beer
- Coconut Water
- Fruit Juices (Apple, Pear, Mango)
- Kombucha
- Malted Drink
- Quinoa Milk
- Rum

- Soy Milk
- Soda
- Tea (Black Tea, Chai Tea, Dandelion Tea, Fennel Tea, Chamomile Tea, Herbal Tea, Oolong Tea)
- Whey Protein
- Wine

Dairy

- Cheese (Cream, Halloumi, Ricotta)
- Custard

- Cream
- Ice Cream/ Gelato
- Kefir

(Cow, Goat, Evaporated Milk, Sheep)

- Sour Cream
- Yogurt

While this may seem like a large list of foods you shouldn't eat, remember that ingredients will affect individuals a little differently. While you should limit the foods listed from above, it is okay to have them every once in a while. The point of this diet is to help reduce symptoms from IBS and bloating. At the end of the day, you are in charge of what you eat and understand how certain foods will make you feel.

Low FODMAP Foods

Fruits

- Ackee
- Breadfruit
- Blueberries
- Bilberries
- Bananas (Unripe)
- Clementine
- Cranberry
- Cantaloupe
- Carambola
- Dragon Fruit
- Guava (Ripe)
- Grapes
- Honeydew
- Kiwi Fruit
- Lime
- Lemon
- Mandarin
- Orange
- Plantain
- Papaya
- Passion Fruit
- Rhubarb
- Raspberry
- Strawberry
- Tangelo
- Tamarind

Vegetables

- Alfalfa
- Butternut Squash
- Brussel Sprouts
- Broccolini
- Broccoli
- Bok Choy
- Beetroot
- Bean Sprouts
- Bamboo Shoots
- Cucumber
- Courgette
- Corn

- Choy Sum
- Cho Cho
- Chives
- Chili
- Chick Peas
- Celery
- Carrots
- Cabbage
- Eggplant
- Fennel
- Ginger
- Green Pepper
- Green Beans
- Kale
- Leek Leaves
- Lentils
- Lettuce
- Olives
- Okra

- Pumpkin
- Peas (Snow)
- Parsnip
- Red Peppers
- Radish
- Sweet Potato
- Swiss Chard
- Sun-Dried Tomatoes
- Squash
- Spinach
- Spaghetti Squash
- Seaweed
- Scallions
- Turnip
- Tomato
- Water Chestnuts
- Yams
- Zucchin

Meat and Poultry

- Beef
- Chicken
- Deli Meats

- Lamb
- Prosciutto
- Pork

118

- Turkey
- Processed Meats

Seafood and Fish

- **Fresh Fish (Cod, Haddock, Salmon, Trout, Tuna, Canned Tuna)**
- **Seafood (Crab, Lobster, Mussels, Oysters, Shrimp)**

Breads, Cereals, Grains, and Nuts

- Bread
  Wheat Free
  Gluten Free
  Potato Flour
  Spelt Sourdough
  Rice
  Oat
  Corn
- Pasta
  Wheat Free
  Gluten Free
- Almonds
- Biscuit (Shortbread)
- Buckwheat (Noodles, Flour)
- Brazil Nuts
- Brown Rice
- Crackers
- Corn Tortillas
- Coconut Milk
- Cornflakes
- Corncakes
- Crispbread
- Corn Flour
- Chips (Plain)
- Mixed Nuts
- Millet
- Macadamia Nuts
- Oatcakes
- Oats
- Oatmeal

- Pretzels
- Potato Flour
- Popcorn
- Polenta
- Pine Nuts
- Pecans
- Rice
  White
  Rice
  Brown
  Basmati
- Rice Krispies

- Rice Flour
- Rice Crackers
- Rice Cakes
- Rice Bran
- Seeds
  Sunflower
  Sesame
  Pumpkin
  Poppy
  Chai
- Tortilla Chips
- Walnuts

Condiments, Sweets, and Sweeteners

- Almond Butter
- Acesulfame K
- Aspartame
- Chocolate
  White
  Milk
  Dark
- Erythritol
- Fish Sauce
- Glycerol
- Glucose

- Golden Syrup
- Jelly
- Ketchup
- Mustard
- Miso Paste
- Mayonnaise
- Marmite
- Marmalade
- Maple Syrup
- Oyster Sauce

- Peanut Butter
- Rice Malt Syrup
- Sucralose (Sugar)
- Stevia
- Sweet and Sour Sauce
- Shrimp Paste
- Saccharine
- Tomato Sauce

- Tamarind
- Vinegar
  Rice Wine Vinegar
  Balsamic Vinegar
  Apple Cider Vinegar
- Worcestershire Sauce
- Wasabi

Drinks

- Alcohol (Wine, Whiskey, Gin, Vodka, Beer)
- Coffee
- Chocolate Powder
- Protein Powder (Whey, Rice, Pea, Egg)
- Soya Milk
- Sugar-Free Soft Drinks
- Water

Dairy/ Eggs

- Butter
- Cheese (Swiss, Ricotta, Parmesan, Mozzarella, Goat, Fetta, Cottage, Cheddar, Camembert, Brie)
- Eggs
- Milk (Rice, Oat, Macadamia, Lactose-free, Hemp, Almond)
- Swiss Cheese
- Soy Protein
- Sorbet
- Tofu
- Tempeh

- Yogurt (Goat, Lactose- free, Greek, Coconut)

Herbs and Spices

- Bay Leaves
- Basil
- Curry Leaves
- Coriander
- Cilantro
- Fenugreek
- Lemongrass
- Mint
- Oregano
- Parsley
- Rosemary
- Sage
- Thyme
- Tarragon
- All Spice
- Black Pepper
- Chili Powder
- Cardamom
- Curry Powder
- Cumin
- Cloves
- Five Spice
- Fennel Seeds
- Nutmeg
- Saffron
- Turmeric
- Avocado Oil
- Coconut Oil
- Canola Oil
- Olive Oil
- Sesame Oil
- Sunflower Oil
- Soy Bean Oil
- Vegetable Oil
- Baking Soda
- Baking Powder
- Cocoa Powder
- Ghee
- Gelatin
- Lard
- Salt
- Yeast

As you can tell from the list from above, there are food choices for all different types of diets. Whether you are vegan, vegetarian, or follow a typical diet, there are plenty of choices for you.

The list from above may seem daunting, but as you learn your own version of the low FODMAP diet, you will be able to put together recipes from the ingredients you are allowed. The key to being successful on this diet is enjoying the foods you are allowed. Luckily in today's market, there are plenty of substitutes for ingredients that may trigger you. As long as you take the time to make this list, you will be able to make your new diet successful.

In the chapter to follow, we will be providing a couple different meal plans for you to follow. There will be a seven-day example vegan diet. Once you have read through this, you can move onto the fourteen-day low FODMAP starter diet. Remember that these are mere suggestions and you can make adjustments as needed.

Chapter 5: Low FODMAP Diet Meal plan

At this point in the book, you hopefully have a better understanding of the foods you can and cannot eat while on the low FODMAP diet. Before we jump into potential meal plans for you to follow, it is time to learn some delicious ingredients.

If you feel nervous about the diet due to the big list of foods to avoid, you absolutely shouldn't! Is your diet going to be different? Yes. However, when you are no longer experiencing diarrhea, constipation, bloating, and the other symptoms from IBS, you will be asking yourself why you didn't start sooner!

As you will find out from the recipes from below, there is a way to stick to your diet and enjoy your meal at the same time. You will find easy to make breakfast, lunch, and dinner recipes. Remember to pay special attention to the ingredients so you can determine if the recipe itself will stick within your own limits.

Low FODMAP Breakfast Recipes:

Small Banana Pancakes

Prep Time: Five Minutes

Cook Time: Twenty Minutes

Servings: Two

Portion: Four Mini Pancakes

Ingredients:

- Dairy-free Spread (Olive Oil) (3 T.)
- Ground Nutmeg (.25 t.)
- Ground Cinnamon (.50 t.)
- Salt (.125 t.)
- Baking Powder (.25 t.)
- Brown Sugar (1 T.)
- Gluten-free All-Purpose Flour (2 T.)
- Egg (2)
- Banana (2 Small, Unripe)

Instructions

1. Begin by heating a medium pan over medium heat before tossing in your dairy-free spread.
2. While this is cooking, go ahead and peel the banana before placing it into a bowl. Mash the banana until it becomes smooth and then add in the egg.
3. Once the egg and banana are mixed well, go ahead and add in the rest of the ingredients. At this point, you should have a mixture that resembles batter.
4. Spoon the mixture into your heated pan and cook the pancakes for a few minutes on each side or until they turn a nice golden color.
5. For extra flavor, try topping the pancakes with your favorite low FODMAP fruit!

Roasted Sausage and Vegetable Breakfast Casserole

Prep Time: Twenty-Five Minutes

Cook Time: Forty-Five Minutes

Servings: Eight

Ingredients:

- Eggs (12)
- Low FODMAP Milk (.50 C.)
- Dried Oregano (.50 t.)
- Salt and Pepper (.25 t.)
- Leek Tips (.50 C.)
- Red Bell Pepper (1)
- Lamb Sausage (1 Package)
- Baby Spinach (2 C.)
- Potato (1)
- Butternut Squash (1)
- Sweet Potato (1)
- Olive Oil (1 T.)

Instructions:

1. Before you begin prepping your food, you will want to preheat your oven to 400 degrees.
2. As your oven heats up, prepare the vegetables from the list above by peeling them and dicing the ingredients into bite-size pieces.

3. Once this is done, place the vegetables on a tray and drizzle them lightly with olive oil or a spread that is allowed on your own low FODMAP diet. Pop them into the heated oven for twenty minutes or until they are soft.

4. While the vegetables are cooking, you can cook your red bell pepper, leek, and sausage in a pan over medium heat. Be sure to cook all of these ingredients through.

5. Now that all of these ingredients are cooked, add in the vegetables to a large casserole dish.

6. In a small bowl, mix together the eggs and add in desired spices. When ready, gently pour the mix over the vegetables already placed in the casserole dish.

7. Place the dish in the oven for thirty minutes or until the eggs are set. This is a great dish to enjoy hot or cold for breakfast!

Blueberry Low FODMAP Smoothie

Prep Time: Five Minutes

Servings: One

Ingredients:

- Lemon Juice (1 t.)
- Maple Syrup (.50 T.)
- Rice Protein Powder (2 t.)
- Frozen Banana (1)
- Ice Cubes (6-10)
- Blueberries (20)
- Vanilla Soy Ice cream (.25 C.)
- Low FODMAP Milk (.50 C.)

Instructions:

1. Place all of the ingredients from above into a blender. Be sure to cut the frozen banana into smaller pieces.
2. Serve right away for a delicious breakfast.

Banana and Oats FODMAP Breakfast Smoothie

Prep Time: Five Minutes

Servings: One

Ingredients:

- Almond Milk (.50 C.)
- Linseeds (1 t.)
- Rolled Oats (1 T.)
- Banana (1)

Instructions:

1. Place all of the ingredients from above into a blender. Be sure you cut the banana into smaller pieces for easier blending.
2. Serve immediately for a filling and healthy meal.

Blueberry, Banana, and Peanut Butter Breakfast Smoothie

Prep Time: Five Minutes

Servings: One

Ingredients:

- Ice Cubes (6-10)
- Low FODMAP Milk (.75 C.)
- Blueberries (.50 C.)
- Peanut Butter (1 T.)
- Banana (.50)

Instructions:

1. Place all of the ingredients from above into a blender and blend until everything is smooth.
2. Serve immediately and enjoy!

Kale, Ginger, and Pineapple Breakfast Smoothie

Prep Time: Five Minutes

Servings: One

Ingredients:

- Ice (1 C.)
- Ginger (.25 T.)
- Kale (1 C.)
- Pineapple (.75 C.)
- Orange (.50)
- Low FODMAP Milk (1 C.)

Instructions:

1. Place all of the ingredients from above into a blender and blend until everything becomes smooth.
2. Serve immediately for a nice, healthy breakfast.

Strawberry and Banana Breakfast Smoothie

Prep: Five Minutes

Servings: One

Ingredients:

- Ice (1 C.)
- Maple Syrup (1 t.)
- Low FODMAP Milk (.75 C.)
- Strawberries (6)
- Banana (1)

Instructions:

1. Toss all of the ingredients into your blender and mix together until smooth.
2. Serve and for an extra treat, try adding some whipped cream!

Low FODMAP Soups and Salads:

Apple, Carrot, and Kale Salad

Prep Time: Ten Minute

Servings: Eight

Portion: .50 C.

Ingredients:

- Salt and Pepper (.25 t.)
- Maple Syrup (1.50 t.)
- Mustard (1 T.)
- Red Wine Vinegar (1.50 T.)
- Olive Oil (3 T.)
- Kale (.50 C.)
- Carrots (3)
- Apple (1 C.)

Instructions:

1. First step, you will want to create your dressing for the salad. You can do this by taking a small bowl and mixing together the maple syrup, mustard, vinegar, and oil. For some extra flavor, season with salt and pepper to taste.

2. Once this is done, take the kale, carrots, and apple and chop into fine, smaller pieces.

3. Finally, dress the salad, toss it a bit, and your meal is ready to be served!

Green Bean, Tomato, and Chicken Salad

Prep Time: Fifteen Minutes

Servings: Four

Portion: .50 C.

Ingredients:

- Lettuce (1 C.)
- Basil Leaves (2 T.)
- Cherry Tomatoes (10)
- Gruyere Cheese (.50 C.)
- Cooked Chicken (1 Lb.)
- Green Beans (.50 C.)

Instructions:

1. To begin, you will want to bring a medium pot of water to a boil. Once the water is boiling, cook your green beans for a few minutes. Once they are tender, drain the water from the pot and run the beans under cold water for a minute.
2. Next, take a large bowl and mix together all of the ingredients from above for a healthy salad.
3. For extra flavor, top your salad with any low FODMAP approved dressings.

Tuna Salad Low FODMAP Style

Prep Time: Five Minutes

Servings: Six

Portion: .50 C.

Ingredients:

- Salt and Pepper (to taste)
- Dried Dill (.50 t.)
- Lemon Juice (1.50 T.)
- Mayonnaise (.50 C.)
- Celery (.50)
- Tuna (2 Cans)

Instructions:

1. Start out by squeezing the liquid out of the tuna.
2. Once you have discarded the tuna, add it into a medium bowl with the vegetables from above.
3. When everything is stirred together, add in the dill, lemon juice, mayonnaise, along with the salt and pepper.
4. This mixture is great for any salad or sandwich!

Low FODMAP Pumpkin Soup

Prep Time: Ten Minutes

Cook Time: Fifteen Minutes

Servings: Six

Portion: 1 C.

Ingredients:

- Lactose-free Half and Half (.75 C.)
- Light Brown Sugar (1 T.)
- Canned Pure Pumpkin (1)
- Vegetable Soup Base (2 T.)
- Water (3 C.)
- Cayenne Pepper (.125 t.)
- Nutmeg (.25 t.)
- Cinnamon (.25 t.)
- Smoked Paprika (.25 t.)
- Scallions (.75 C.)
- Olive Oil (1 T.)
- Unsalted Butter (2 T.)
- Salt and Pepper to taste

Instructions:

1. To begin, you will want to heat up a medium sized pot over a low to medium heat. Once the pot is warm, you can add in your oil

and butter until it begins to sizzle.

2. When the butter and oil are warm, add in your spices with the scallions and cook until they are soft.

3. Once this happens, add in the soup and water. Be sure to mix everything together before you add in the salt, brown sugar, and the canned pumpkin.

4. Now that these ingredients are placed in the pot, lower your heat and allow these to simmer for ten minutes or so. Feel free to stir every once in a while, to assure the ingredients are blended well.

5. Now, remove the soup from the heat and add in your half and half. Once the soup is cool, you can place the mixture into a blender and blend until it is smooth.

6. For extra flavor, season the soup with salt and pepper to taste.

Quinoa and Turkey Meatball Soup

Prep Time: Fifteen Minutes

Cook Time: Twenty Minutes

Servings: Eight

Portion: 1 C.

Ingredients:

- Collard Greens (5 C.)
- Celery (.50)
- Leek Tips (1 C.)
- Olive Oil (2 T.)
- Egg (1)
- Dried Basil (2 T.)
- Parsley (2 T.)
- Cooked Quinoa (.50 C.)
- Ground Turkey (1 Lb.)
- Turkey Stock (10 C.)
- Salt and Pepper to taste

Instructions:

1. To start out, you are going to want to make your meatballs for the soup. You will do this by taking a large mixing bowl and combine the egg, parsley, basil, quinoa, and turkey together. Gently take the mixture in your hands and form one inch balls.

2. Next, take a medium pan over medium heat and cook the turkey meatballs in olive oil for a few minutes. Be sure to flip the balls over so that they are a nice golden-brown color all around.

3. Now that these are done, take a large pot over medium heat and

add in a tablespoon of oil. Once the oil is sizzling, you can add in the leek and celery. Sauté these two ingredients for a minute before adding in the collard greens and stock.

4. When all of the ingredients are cooked, add in the meatballs and allow this mixture to simmer over a low heat for eight to ten minutes.

5. Remove the soup from the heat and allow to cool slightly before serving.

Mixed Vegetable, Bean and Pasta Soup

Prep Time: Fifteen Minutes

Cook Time: Thirty Minutes

Servings: Fourteen

Portion: .75 C.

Ingredients:

- Gluten-free Pasta (1 C.)

- Dried Thyme (1 t.)

- Smoked Paprika (1 t.)

- Dried Basil (1 t.)

- Zucchini (1)

- Squash (1)

- Bok Choy (2 C.)

- Carrots (3)

- Kale (1 C.)

- Red Potatoes (1 C.)

- Butternut Squash (1 C.)

- Crushed Tomatoes (1 Can)

- Water (8 C.)

- Leek Tips (.25 C.)

- Scallions (.75 C.)

- Olive Oil (2 T.)

- Salt and Pepper to taste

Instructions:

1. To start, you will want to take a large pot and begin to heat it over medium heat with the olive oil placed in the bottom.

2. Once the olive oil is sizzling, add in the leeks and scallions and allow them to cook until they become soft.

3. When these are ready, add in your prepared zucchini, squash, Bok choy, carrots, kale, potatoes, chickpeas, canned tomatoes, and the water. Season as desired and place the top on the pot.

4. Bring all of the ingredients from above to a boil and then turn the heat down to allow everything to simmer for at least thirty minutes. By the end, all of the vegetables should be tender.

5. While the soup cooks, you can cook the gluten-free pasta in another pot so by the end, you can combine everything and have a healthy meal!

Vegan Options:

Low FODMAP Coconut and Banana Breakfast Cookie

Prep Time: Ten Minutes

Cook Time: Twenty Minutes

Servings: Ten

Portion: One

Ingredients:

- Vanilla Extract (1 t.)
- Vegetable Oil (.25 C.)
- Maple Syrup (.25 C.)
- Banana (1)
- Baking Powder (.50 t.)
- Cinnamon (1 t.)
- Ground Flax Seeds (2 T.)
- Chia Seeds (2 T.)
- Unsweetened Coconut Flakes (.50 C.)
- Banana Chips (.50 C.)
- Gluten-free All-purpose Flour (.50 C.)
- Old-fashioned Oats (1 C.)

Instructions:

1. You will want to begin by heating your oven to 325 degrees.

2. While the oven heats up, take a medium bowl and mix together the baking powder, cinnamon, flax seeds, chia seeds, coconut flakes, banana chips, flour, and oats altogether.

3. In another bowl, mix together a mashed banana, vanilla, vegetable oil, and pale syrup. When both bowls are well combined, you can mix them together and begin to create your dough.

4. Next, take a greased cookie sheet and lay out balls of dough to create your cookies. When this is done, pop the cookie sheet in the oven for twenty minutes.

5. When the time is up, remove the cookies, allow to cool, and enjoy!

Lemon and Garlic Roasted Zucchini

Prep Time: Five Minutes

Cook Time: Twenty Minutes

Servings: Twelve

Portion: 1 C.

Ingredients:

- Olive Oil (1.50 T.)
- Zucchini (2)
- Lemon Zest (2 T.)
- Salt and Pepper to taste

Instructions:

1. You can begin by heating your oven to 425 degrees.
2. While this warms up, slice your zucchini into thin slices and place in a bowl with the lemon zest and olive oil. Assure it is covered completely before seasoning with salt and pepper.
3. Place the zucchini on a greased sheet pan and cook for twenty minutes.

Rainbow Low FODMAP Slaw

Prep Time: Ten Minutes

Servings: Twenty

Portion: 1 C.

Ingredients:

- Pomegranate Seeds (.50 C.)
- Carrots (3)
- Kale (1 C.)
- Red Cabbage (1 C.)
- Green Cabbage (1 C.)
- Lactose-free Yogurt (.50 C.)
- Dijon Mustard (1 t.)
- Sugar (2 T.)
- Apple Cider Vinegar (.25 C.)
- Canola Oil (.50 C.)

Ingredients:

1. Start out by creating your dressing for the slaw. You can do this in a small bowl, mix together the canola oil, apple cider vinegar, Dijon mustard, sugar, yogurt, and a little bit of salt.
2. In another bowl, toss together the different cabbage with the carrots and the kale.
3. Gently drizzle the dressing over the kale, and you have a delicious slaw that is full of color and flavor!

Vegan Roasted Red Pepper Farfalle

Prep Time: Ten Minutes

Cook Time: Ten Minutes

Servings: Four

Portion: 1 C.

Ingredients:

- Capers (.25 C.)
- Parsley (.75 C.)
- Olive Oil (.25 C.)
- Roasted Red Peppers (1 Jar)
- Gluten-free Farfalle Pasta (2 C.)

Instructions:

1. You can start this recipe by cooking your pasta according to the instructions on the side of the box.
2. Once the pasta is cooked through, drain the water and then place the pasta back into the pot.
3. Toss in the oil, parsley, roasted red peppers, and capers to the mixture.
4. Mix everything together and season with salt and pepper for extra flavor.

As you can tell, you can follow the low FODMAP diet and still enjoy delicious foods. While these are only some of the many recipes you can follow on your diet, there are plenty of resources out there to provide you with even more! With these resources in hand, we will now go over a simple seven and fourteen-day meal plan that is easy to follow.

With a limited food choices, you may be thinking to yourself that you are going to get bored quick. When it comes to a new diet, it is all about your frame of mind. On one hand, you could think negatively about it and return to your old eating habits. With choice, comes consequence. When you eat the foods that trigger you, you are going to feel lousy. Why make that choice when you can choose to eat healthy and feel better? Below, we will provide some simple meals for you to consider until you feel confident enough to create your own recipes

Breakfast Meal Plan Ideas:

- Eggs- Hard-boiled, over easy, or even scrambled. There are many ways to enjoy eggs!
- Lactose-Free Yogurt with any low FODMAP fruit
- Gluten-free Muffins
- Gluten-free French toast
- Gluten-free Oatmeal with cinnamon
- Rice Cereal with low FODMAP fruit
- Ground Turkey
- Smoothie with low FODMAP fruit

Lunch Meal Plan Ideas:

- Gluten-free Bread with Deli Meat and Cheese

- Chicken Noodle Soup
- Quinoa Bowl with low FODMAP Veggies or Grilled Chicken
- Salad
- Baked Potato with Lactose-free Butter

Dinner Meal Plan Ideas:
- Stir-Fried Rice
- Tacos
- Gluten-Free Pizza
- Grilled Chicken Salad
- Steak with Fresh Low FODMAP Vegetables
- Grilled Chicken with White Rice
- Rice Pasta with Marinara
- Snack Meal Plan Ideas:
- Rice Cakes with Peanut Butter
- Baby Carrots
- Lactose-free Yogurt
- Unripe Banana
- Unsalted Peanuts
- Pop Chips
- Gluten-free Pretzels
- Crackers with Cheese
- Hard-Boiled Egg

14- Day Meal Plan

Week One:

Meal	Monday	Tues.	Wed.	Thurs.	Friday
BFast	Small Banana Pancakes	Blueberry Smoothie	Roasted Sausage and Vegetable Breakfast Casserole	Strawberry and Banana Breakfast Smoothie	Banana and Oats FODMAP Breakfast Smoothie
Lunch	Apple, Carrot, and Kale Salad	Mixed Vegetable, Bean, and Pasta Soup	Low FODMAP Pumpkin Soup	Tuna Salad Low FODMAP Style	Quinoa and Turkey Meatball Soup
Dinner	Low FODMAP Veggie Latkes	Steak with Lemon and Garlic Roasted Zucchini	Left Over Mixed Vegetable, Bean, and Pasta Soup	Vegan Roasted Red Pepper Farfalle	Salad with Grilled Chicken and Homemade Dressing

Meal	Saturday	Sunday
Breakfast	Eggs and low FODMAP fruit	Rice Cereal with low FODMAP fruit
Lunch	Chicken Noodle Soup	Baked Potato with Lactose-free Butter
Dinner	Stir-Fried Rice	Gluten-free Pizza

Week Two:

Meal	Monday	Tuesday	Wednesday	Thursday	Friday
Breakfast	Gluten-free French Toast	Rice Cereal with low FODMAP fruit	Lactose-free Yogurt with low FODMAP fruit	Blueberry Smoothie	Small Banana Pancakes
Lunch	Quinoa Bowl	Salad with Approved Dressing	Gluten-free Sandwich with Deli Meat and Cheese	Mixed Vegetable, Bean, and Pasta Soup	Tuna Salad on Gluten-free Bread
Dinner	Grilled Chicken with White Rice	Grilled Chicken Salad	Salad with Approved Dressing	Gluten-free Tacos	Stir-Fried Rice

Meal	Saturday	Sunday
Breakfast	Smoothie with low FODMAP fruit	Lactose-free Yogurt with low FODMAP fruit
Lunch	Rice Pasta with Marinara	Chicken Noodle Soup
Dinner	Baked Potato with Lactose-free Butter	Grilled Chicken Salad

Vegan 7-Day Meal Plan

Meal	Monday	Tuesday	Wednesday	Thursday	Friday
Breakfast	Coconut Yogurt with Chia Seeds	Rice Cakes with Peanut Butter	Corn Flakes with Almond Milk	Gluten-free Bread with Almond Butter	Unripe Banana with Coconut Yogurt
Lunch	Lemon and Garlic Roasted Zucchini	Rainbow Low FODMAP Slaw with Gluten-free Bread	Vegan Roasted Red Pepper Farfalle	Low FODMAP Coconut and Banana Cookie with Coconut Yogurt	Salad with Approved Dressing
Dinner	Low FODMAP Veggie Latkes	Gluten-free Pasta with Approved Sauce	Plain Tempeh with low FODMAP Veggie of choice	Plain Tofu with Rice Noodles	Gluten-free Pizza with Soy Cheese

Meal	Saturday	Sunday
Breakfast	Blueberry Smoothie with Coconut Milk	Banana and Oat Smoothie with Coconut Milk
Lunch	Plain Tofu with Soba Noodles	Plain Tempeh with Gluten-free Pasta
Dinner	Grilled Cabbage Soup	Baked Brussel Sprouts with Plain Tofu

Chapter 6: Low FODMAP diet tips and tricks for success

Starting a new diet can be scary. As we said before, your frame of mind is going to be incredibly important. It is vital you think about your why when making food choices. Each meal, we have a chance to better our health; all it takes is a little thought behind each decision.

Of course, we want to see you succeed with your diet. Below, you will find a number of tips and tricks that have helped other clients on the low FODMAP diet. While some may work for you, others may not. You must adjust the low FODMAP diet to match your desired lifestyle so you not only stick with it but can enjoy it at the same time!

1.   Read the Label

Reading the labels on packaged foods is going to be vital for the success of your diet. Unfortunately, many high FODMAP ingredients can have very confusing names. We suggest carrying a list of additives to avoid until you learn them by heart. When you are more aware, you can avoid the high FODMAP ingredients.

2.   Water-soluble

In general, low FODMAP foods are going to be water-soluble, but this does not mean they are fat-soluble. If you are cooking a soup with an onion, you will want to take the onion out. Instead, try using onion-infused oils for the taste. It is quick fix that may help with your IBS triggers.

1.   High Fructose Corn Syrup

High Fructose Corn Syrup is in everything. Again, it will be important that you learn how to read food labels so you will be able to avoid this mistake. This ingredient is in a number of foods including energy bars, juices, mayonnaise, frozen meals, and even popcorn. Check the label before you put anything into your shopping cart.

2.   Fiber

If you pick up a product and it seems to have a high serving f fiber, you can assume it is due to a high FODMAP additive. Try to avoid any products that boast about their fiber; it's a trap! Any fiber additives will more than likely trigger your GI issues.

3.   Onion and Garlic Powder

When it comes to choosing out your spices, pay special attention to the labels. You will want to avoid onion and garlic powders as they contain high FODMAPs. Luckily, there are plenty of delicious low FODMAP approved spices as you can find in the chapter from above.

4.   "Natural"

If you find any frozen foods, brothers, or savory soups that claim they have "natural" flavors, go ahead and check out the label. You can assume that they contain garlic and onion, very popular IBS triggers. You will want to try your best to avoid these additives to your meals.

5.   "Healthy"

As much as we would like to trust when products claim they are healthy, this does not equate to low FODMAP approved. Foods like asparagus and apples are supposedly "healthy" for you, but they can trigger IBS symptoms. As you go through the elimination process, you will learn just what you can and cannot eat and make the decision if something is healthy for you.

6. Beverages

Often times, people forget that beverages can contain FODMAPs. You will want to pay special attention to what you are putting into your body. If you ever have questions, feel free to refer to our lists in the chapter from above. Just because a beverage claims it has no net carbohydrates, this does not mean they aren't high in FODMAPs.

7. Portion Control

While you are on the low FODMAP diet, portion is going to be key to success. When you are reading labels, you will always want to pay special attention to portion size. While a low FODMAP diet is approved, a bigger portion may still trigger IBS symptoms. You will want to try your best to be mindful of portion control.

8. Learn About Yourself

As you start this diet, you will want to spend plenty of time on the elimination phase. The more you test, the more you will be able to figure out what foods you can and cannot eat. When you have more to choose from, you will be able to get more creative with your recipes. At the end of the day, only you know what is best for you. When you learn yourself, the diet will become that much easier.

9. Food and Meal Journal

Your food and meal journal are going to be an important tool for your low FODMAP diet journey. By keeping track of the foods, you can and cannot eat; it will make it easier when you go back to check out your history. We eat so many different types of foods through the day; it can be hard to remember which foods trigger you. By keeping a journal, it leaves little room for mistakes.

10. Use Your Fridge

If you are trying your best to stick with the low FODMAP diet, why leave anything up to chance? Do yourself a favor and take the time to remove any high FODMAP foods in your house. By keeping your pantry and fridge stocked with the low FODMAP foods you need, it deletes any temptations you may have in the house.

11. Have A Backup Plan

Dieting is hard, especially when you are first starting out. When you are planning out your meals, it is possible to miss one here or there. Try to stack your freezer with low FODMAP meals so you can cook them in a few minutes. When your first plan falls through, you will always have a backup. It is a win-win situation!

Chapter 7: Low FODMAP diet FAQ

As we are nearing our time together, hopefully, you are feeling better about starting the low FODMAP diet. While you have learned a lot about the diet, feel free to check back whenever you have a question about the diet. Whether you need a refresher on the benefits of the diet or a reminder of which foods you can and cannot eat, you will be able to find the information here easily.

To finish off, we will hopefully be able to answer any further questions you have about the low FODMAP diet. Simply remember that this diet is going to be specifically tailored up to you. Being diagnosed with IBS or other GI tract issues is not the end of the world. It will take some extra effort, but when your symptoms and discomfort are relieved, you will be thankful you made the choice to start the low FODMAP diet. For now, it is time to answer some more popular questions you may still have.

Q. I am following the low FODMAP diet and still experiencing symptoms, is this the right diet for me?

A. The answer could be yes and no. If you are following the diet and still find yourself with symptoms of IBS, there may be another culprit in your diet. Remember to keep a food diet with you at all times so you can find any triggers you may be missing.

Q. Can I follow the low FODMAP diet as a vegetarian?

A. Absolutely! You can follow this diet whether you are vegan or vegetarian, it will just take a little extra work. You will find in the chapters before there are plenty of choices, so long as the allowed foods are not triggers for your own body. Some good sources of protein for this diet would be chickpeas, tofu, tempeh, and more. If there is a will, there is always a way!

Q. How do I make sure I'm getting enough Fiber?

A. This is one of the bigger concerns for those following the low FODMAP diet, especially if constipation is an issue. Luckily, several good low FODMAP sources can help you keep your fiber intake up. These include chia seeds, brown rice, flax seeds, kiwi, oranges, white potato, rice bran and more. Check out the list provided in the fourth chapter for a longer list.

Q. Should I eat Larger or Smaller Meals on the Low FODMAP Diet?

A. In general, you should try to eat three main meals through your day and to snacks between these meals. If you are still hungry, you can always add in another snack. Remember that portion control is going to be vital while you are on the low FODMAP diet, so this is something you will want to keep in mind when plating your meals.

Q. What is the rule with fats and oils on the Low FODMAP Diet?

A. As a general rule, there are plenty of fats and oils that are low in FODMAPs. However, anything in excess can trigger IBS symptoms. You will want to be especially aware of any condiments or sauces that are oil-based such as salad dressings. Most of the time, these also include high FODMAPs like garlic. Remember to always read the labels before consuming anything. You can also refer to our extended grocery list to see which fats and oils are allowed on the diet.

Q. Can I eat meat on the low FODMAP Diet?

A. Yes and No. Some sources of animal protein such as fish and chicken are low in FODMAPs. However, if the meat is prepared already, you will want to avoid any additives that may trigger your symptoms. If you have any further questions, please refer to the food lists from the chapter above.

Q. What happens if I break my diet?

A. While the aim of the diet is to stick to it as much as possible, mistakes and slip ups will happen. Overall, you will want to achieve control over any symptoms you may be having. If you slip up, expect to experience the IBS symptoms. As long as you return to your diet, you will most likely be able to improve them in a few days.

Q. Is this a lifestyle?

A. No, the low FODMAP diet is not meant to last for a lifetime. The aim of the diet is to help heal your gut over a controlled period of time. This diet should only be followed for two to six weeks. After this, you can begin to introduce food back into your diet. This will change depending on each individual.

PART IV

For all the meat lovers out there, the next diet that we are going to discuss is known as the carnivore diet.

Chapter 1: What Is Carnivore Diet?

The Carnivore Diet is the all-new trendy diet that expects its followers to go on a meat-only way of lifestyle. This diet completely goes against nutritional stereotyping. If someone asked you to replace that bowl of meat with vegetable oils and carbs, you probably have been misled!

This diet has won favor for several reasons and is, of course, not a fluke. If you find it attractive to sink in a carnivore way of eating, you have come to the right place.

The carnivore diet is the one that entirely revolves around a meat-based pattern of eating. It is one extreme diet that restricts you from eating plant-based foods and strictly opposes carbohydrate consumption. It might sound crazy, but there are people including Shawn Baker (the creator of this diet) who have normalized this fact that carbohydrate is a non-essential macronutrient, and there is no harm in cutting them down from the menu.

It is a zero-carb diet that altogether emphasizes consuming meat. Scientists consider this diet the best nutrition source for human beings, cutting out the chatter from plant toxins.

There are more anecdotes and testimonials than research backed up with science. This no-carb diet has also been a second option to those who have either failed to carry on with Paleo and Keto diet or have faced other severe consequences after following them. This is a bold claim and has a lot to unpack. Let's dive deep into the depth of its science.

Researchers have carried out several studies throughout the Earth that has proved its benefits on humankind.

- **Removes the Inflammatory Vegetables -** If you have been suffering from an autoimmune disorder or a damaged gut, this might be of concern.

  Almost every vegetable has some kind of toxin in it. Brussels Sprouts, broccoli, cauliflower have sulforaphane that causes hypothyroidism and damage to health. Nightshades damage carb and fat metabolism. Polyphenols cause DNA damage. Lectins cause leaky guts. Reservatrol can inhibit androgen precursors. Spinach has oxalates that may result in kidney stones, and the list goes on. Choosing a carnivore diet can be a game-changing plan for such people and others who are still being manipulated by conventional nutritional advice.

- **It Increases Cholesterol –** You all must have heard about bad and good cholesterol. Cholesterol plays a negative role when it is oxidized or damaged and gets trapped in the artery walls. LDL cholesterol, even though it is given the tag of the 'bad' cholesterol, protects your body from diseases and does not cause them. They bind to the pathogens allowing the immune system to expel them.

During inflammation, the body uses LDL as a protective mechanism. So, people with heart diseases have high LDL levels because it binds to the pathogens, getting rid of the damage ensuring that it does not spread. It is, in fact, the inflammation that causes heart diseases.

- **It Increases the Nutrient Density** - Animal-based foods have the most bioavailable form of nutrients that play a crucial role from growth to brain function. While you have been a fiber-freak, you might just have missed out on the essential nutrients. Vegetarians have a deficiency in Vitamin B12 and Iron. Americans have Vitamin D deficiency, and women have Calcium deficiency, while Zinc is a deficient nutrient worldwide.

The brain requires micronutrients, and it is animals that mostly provide this. Zinc and iron are vital nutrients that help brain growth, dopamine transport, and serotonin synthesis.

- **It Reverses Insulin Resistance -** The best thing you could do to your health is reverse the insulin resistance. It is a problem where your body's cells become unresponsive to insulin action and therefore refuse to stuff the cells with more energy, leading to a rise in insulin level. It occurs due to excess carbohydrates and fat that shut off the process of burning fat, causing the fat to be stored without being used directly. The carnivore diet can be a solution to this!

- **Weight loss -** Since protein-based foods are satiating, they allow you to stay distracted from eating by making you feel fuller. By ingesting protein, the primary energy source is shifted from carbohydrate to fat. It is similar to ketosis (adapted to fat consumption), where you can use your body fat instead of carbohydrate.

Although it is as simple as a diet can be, the initial weeks can be hard. Here are the things that you can incorporate to get through the changeover conveniently:

- Before starting with the diet, you can get your blood test done since the metabolic needs vary with every individual.

- You might feel like giving up at some point, start getting headaches, and experience fatigue. It is normal as your body will be getting used to using energy from fat rather than carbohydrates.

- Your eating desire might fluctuate. You will get adjusted to this form of eating after one week.

Chapter 2: Recipes for Tasty Appetizers

If you are a meat lover and want to start the carnivore diet, here are some recipes for you to follow.

Oven-Baked Chicken Wings

Total Prep & Cooking Time: 1 hour 5 minutes

Yields: 8 servings

Nutrition Facts: Calories: 348 | Carbs: 1g | Protein: 25g | Fat: 27g | Fiber: 1g

Ingredients:

- Half cup of grated Parmesan
- Four pounds of chicken wings
- One tsp. of salt
- One tbsp. of parsley
- A quarter cup of grass-fed butter
- Half tsp. of black pepper (ground)

Method:

1. First of all, the oven needs to be preheated to 180 degrees Celsius or 350 degrees Fahrenheit.

2. Take a parchment paper for lining the baking sheet.

3. Now, you need to take a shallow bowl or dish for melting the butter.

4. In another clean bowl, mix parsley, pepper, Parmesan cheese, and salt.

5. Once the herb and cheese mixture is ready, dip the chicken wings in the bowl of melted butter one by one. After dipping, roll the wings in the mixture.

6. Arrange all the wings properly on top of the baking sheet.

7. Bake for an hour.

8. Take out the baked chicken wings from the oven and serve them warm.

Steak Nuggets

Total Prep & Cooking Time: 55-60 minutes

Yields: 4 servings

Nutrition Facts: Calories: 350 | Carbs: 1g | Protein: 40g | Fat: 20g | Fiber: 2g

Ingredients:

- One pound of beefsteak or venison steak (cut it into chunks)
- Palm or lard oil (needed for frying)
- One large-sized egg

For Keto Breading,

- Half cup each of
  - Pork panko
  - Parmesan cheese (grated)
- Half tsp. of seasoned salt (homemade)

For Chipotle Ranch Dip,

- A quarter cup each of
  - Organic cultured cream (sour)
  - Mayonnaise
- More than one tsp. of chipotle paste (for taste)
- A quarter medium-sized lime (juiced)

Method:

1. For preparing the Chipotle Ranch Dip, you need to combine all the ingredients and mix properly. Use either more or less chipotle paste in accordance to your taste preference. Refrigerate the dip before serving for a minimum of thirty minutes. You may store the dip for nearly a week.

2. Take a large-sized bowl and combine parmesan cheese, seasoned salt, and pork panko. Set aside after mixing evenly.

3. Now, beat one egg. Place the breading mix in one bowl and beaten egg in another.

4. Dip the steak chunks first in egg and then in the breading mix. Then, place them on a plate or sheet pan lined with wax paper.

5. Before frying, freeze the raw breaded steak bites for half an hour. By doing so, the breading won't lift at the time of frying.

6. Heat the lard to 325 degrees Fahrenheit. Fry the chilled or frozen steak nuggets for nearly two to three minutes until you get the brown color.

7. Keep the fried nuggets on a plate lined with a paper towel. Sprinkle a pinch of salt. Serve hot along with Chipotle Ranch.

Grilled Shrimp

Total Prep & Cooking Time: 10 minutes

Yields: 4 servings

Nutrition Facts: Calories: 102 | Carbs: 1g | Protein: 28g | Fat: 3g | Fiber: 0g

Ingredients:

For grilling,

- One lb. of shrimp
- One tbsp. of lemon juice (freshly squeezed)

- Two tbsps. of olive oil (extra-virgin)
- For frying – vegetable oil or canola oil

For the shrimp seasoning,

- Half a tsp. of cayenne pepper
- One tsp. each of
  - Italian seasoning
  - Kosher salt
  - Garlic powder

Method:

1. You have to preheat your grill for this recipe on high.

2. Take a mixing bowl of large size, add all the ingredients of the seasoning in it and mix them well. Drizzle the lemon juice and olive oil into the mixture and keep stirring until you get a paste.

3. Add the shrimp into the bowl of seasonings and keep tossing so that all the pieces are evenly coated. Take the shrimp pieces and thread them onto wooden skewers.

4. Coat your grill with canola oil. You have to grill the shrimp for about three minutes for each side until they become opaque and pink.

5. Serve and enjoy!

Notes: You can store the grilled shrimp in the refrigerator for up to three days if you want to, but for the best flavor, you should consume it on the same day.

Roasted Bone Marrow

Total Prep & Cooking Time: 20 minutes

Yields: 2 servings

Nutrition Facts: Calories: 440 | Carbs: 0g | Protein: 4g | Fat: 48g | Fiber: 0g

Ingredients:

- To season – freshly ground black pepper and sea salt flakes
- Four bone marrow halves

Method:

1. Set the temperature of your oven to 350 degrees F and preheat.

2. Take a baking tray with deep sides and then place the bone marrow pieces in it.

3. Bake the bone marrow for half an hour until they become crispy and golden brown in color. The fat that is present in excess should have rendered off by now.

4. Season with black pepper and sea salt flakes.

5. You can spread the marrow separately on top of steaks, or you can serve the bone marrow as an appetizer.

Bacon-Wrapped Chicken Bites

Total Prep & Cooking Time: 30 minutes

Yields: 4 servings

Nutrition Facts: Calories: 230 | Carbs: 5g | Protein: 22g | Fat: 13g | Fiber: 1g

Ingredients:

- Three tbsps. of garlic powder
- Eight slices of thin bacon (slice them into one-third pieces)
- One chicken breast (large-sized, cut into bite-sized pieces)

Method:

1. Set the temperature of the oven to 400 degrees F and use aluminum foil to line the baking tray—Preheat the oven.

2. In a bowl, add the garlic powder. Take each chicken piece and dip it into the garlic powder.

3. Now, take each short piece of bacon and wrap it around the piece of chicken. Keep these prepared chicken pieces on the baking tray. Make sure they are not touching each other.

4. Bake the preparation for half an hour, and by the end of it, the bacon should turn crispy. After about fifteen minutes through pieces, turn the pieces over.

Salami Egg Muffins

Total Prep & Cooking Time: 25 minutes

Yields: 12 servings

Nutrition Facts: Calories: 142 | Carbs: 1g | Protein: 12g | Fat: 10g | Fiber: 0g

Ingredients:

- Four eggs (large-sized)
- Twenty slices of salami (uncured)
- Half a tsp. of kosher salt
- A quarter tsp. of black pepper
- Olive oil

Method:

1. Set the temperature of the oven to 350 degrees F and preheat. Take ramekins of four ounces each and spray them with olive oil. Then, place these ramekins on the baking sheet.

2. On the bottom of each ramekin, place one slice of salami and then on the sides, arrange four slices so that they are overlapping each other.

3. In this way, you will get a basket of salami, and in the middle of the basket, break one egg. Form four such baskets. Season the baskets with pepper and salt.

4. Bake the prepared salami baskets for twenty minutes, and by that time, they should be set.

5. Around the edges of the muffins, run a knife, and the muffins will get released. Serve and enjoy!

3-Ingredients Scotch Eggs

Total Prep & Cooking Time: 40 minutes

Yields: 12 servings

Nutrition Facts: Calories: 270 | Carbs: 1g | Protein: 19g | Fat: 20g | Fiber: 5g

Ingredients:

- Twelve large-sized boiled eggs
- Two pounds of chicken sausage or ground beef
- Two tsps. of salt

Method:

1. Preheating the oven to a temperature of 175 degrees Celsius or 350 degrees Fahrenheit is the first step for preparing such a delicious appetizer.

2. Line two baking sheets (small rimmed) with a parchment paper.

3. Take a large-sized bowl and combine chicken or beef and salt. Mix both the ingredients together by using your hands and then form twelve meatballs with it. Press the meatballs flat after placing them on top of the lined sheets.

4. Now, place each boiled egg inside each circle of flattened meat. After placing the eggs, start wrapping the meat nicely around the eggs. You are not supposed to leave any holes or gaps.

5. You need to bake for nearly fifteen minutes. Flip over as soon as the top looks cooked and again bake for ten minutes. If you want a crispy shell, then finish it under a broiler for approximately five minutes.

Notes: *If you are willing to enhance the taste, then you may feel free to add any of your favorite herbs, such as garlic or rosemary powder. Add one tsp. of your preferred herb into the meat just before wrapping the eggs. Hard-boiled eggs are better in this case as it is difficult to peel the soft boiled eggs.*

Chapter 3: Quick and Easy Everyday Recipes

Carnivore Waffles

Total Prep & Cooking Time: 6 minutes

Yields: 1 serving

Nutrition Facts: Calories: 274 | Carbs: 1g | Protein: 23.6g | Fat: 20.2g | Fiber: 0.8g

Ingredients:

- One-third cup of mozzarella cheese
- One egg
- Half cup of pork rinds (ground)
- A pinch of salt

Method:

1. For preparing the carnivore waffles, all you need is a waffle maker. First of all, preheat your waffle maker (medium-high heat).

2. Take a medium-sized bowl and whisk the pork rinds, cheese, and salt together.

3. Once you are done with the whisking part, pour the already prepared waffle mixture in the middle of the waffle maker's iron.

4. Close it and allow it to cook for three to five minutes. Or, you may cook until the waffle gets an attractive golden brown color.

5. Now, remove the cooked waffle and serve hot.

Notes: *The carnivore waffle will turn out to be more delicious if you place a cube of butter or runny egg on top of it. Greasing the waffle maker is not required before you start cooking waffles.*

Chicken Bacon Pancakes

Total Prep & Cooking Time: 20 minutes

Yields: 4 servings

Nutrition Facts: Calories: 444 | Carbs: 0g | Protein: 33g | Fat: 34g | Fiber: 0g

Ingredients:

- Four bacon slices
- Two chicken breasts
- Two tbsps. of coconut oil
- Four eggs (medium-sized, whisked)

Method:

1. First, you need to add all the ingredients to the bowl of the food processor except for the oil and then process everything together to form a smooth mixture.

2. After that, take your frying pan, and coconut oil to it.

3. From the batter that you just made, form four pancakes.

4. Fry these pancakes until they are set and properly cooked. Do the same with the rest of the batter.

Garlic Cilantro Salmon

Total Prep & Cooking Time: 25 minutes

Yields: 4 servings

Nutrition Facts: Calories: 294 | Carbs: 1g | Protein: 38.9g | Fat: 14.2g | Fiber: 0g

Ingredients:

- One lemon
- One fillet of salmon (large)
- A quarter cup of cilantro leaves (freshly chopped)
- Four garlic cloves (minced)
- One tablespoon of butter (optional)
- To taste – freshly ground black pepper and kosher salt

Method:

1. Set the temperature of the oven to 400 degrees F and preheat. Take a baking sheet and line it with foil. Place the fillets of salmon on it. You don't have to grease the foil.

2. Sprinkle the juice of one lemon over the fillet of salmon. Spread cilantro and garlic on top of the fillets evenly and season with pepper and salt. If you want to use butter, then you have to place thin slices on top of the salmon fillet at this stage.

3. Now, place the salmon along with the foil in the oven and bake for about seven minutes.

4. Set broil settings and cook the salmon for an additional seven minutes. The top part should become crispy.

5. Use a flat spatula to remove the salmon from the oven. Separate the skin from the fish and serve!

Mustard-Seared Bacon Burgers

Total Prep & Cooking Time: 30 minutes

Yields: 6 servings

Nutrition Facts: Calories: 525 | Carbs: 3g | Protein: 22g | Fat: 45g | Fiber: 4g

Ingredients:

- 1.5 pounds of ground beef
- Four ounces of diced bacon
- Six tbsps. of yellow mustard
- To taste – salt and pepper

For the toppings,

- One tomato (properly diced)
- Half a red onion (diced)
- One avocado (thinly sliced)

For the sauce,

- Two tsps. of yellow mustard
- One tsp. of tomato paste
- A quarter cup of mayo

Method:

1. Take a pan and cook the bacon in it until it becomes crispy. You have to keep the grease of the bacon separately so that it can be used later. Then, take the bacon bits and keep them in a bowl along with the ground beef. Use pepper and salt to season them.

2. You will be able to form six patties from the mixture.

3. Now, you have to fry these burger patties on high flame so that they can get a great color. If you want, you can also choose to grill them.

4. Each patty will then have to be coated with one tbsp. of mustard and then, place the patty on the pan with the mustard-side facing down. Sear the patties one by one.

5. Take another bowl in which you can mix all the ingredients of the sauce together.

6. Each burger patty will have to be coated with sauce, and then, you can top them with slices of avocado, tomato, and onions.

Crockpot Shredded Chicken

Total Prep & Cooking Time: 6 hours

Yields: 8 servings

Nutrition Facts: Calories: 201 | Carbs: 1g | Protein: 24g | Fat: 10g | Fiber: 0g

Ingredients:

- Four garlic cloves
- Four chicken breasts
- One cup of chicken broth
- Half an onion (sliced)
- One tbsp. of Italian seasoning
- To taste – Salt and pepper

Method:

1. Take all the ingredients and add them to the crockpot.

2. Cook them for about six hours on low.

3. Use forks to shred the meat.

4. You can enjoy the shredded chicken with a variety of dishes like sautés, lettuce wraps, salads, or even soups.

Chapter 4: Weekend Dinner Recipes

Organ Meat Pie

Total Prep & Cooking Time: 20 minutes

Yields: 4 servings

Nutrition Facts: Calories: 412 | Carbs: 2g | Protein: 35g | Fat: 28g | Fiber: 4.2g

Ingredients:

- Half pound each of
  - Beef liver (ground)
  - Beef heart (ground)
  - Ground beef
- Three eggs
- Butter, ghee or Homemade Tallow or any melted cooking fat
- Salt (as required)

Method:

1. The oven needs to be preheated to 175 degrees Celsius or 350 degrees Fahrenheit.

2. Take a mixing bowl: mix ground beef, beef heart, and beef liver along with eggs and cooking fat of your choice. Lastly, add salt into the mixture.

3. Now, take a pie plate of nine inches and grease it lightly. Pour the mixture into the pie plate evenly.

4. Bake it for nearly fifteen to twenty minutes. Or, you may bake until the egg is totally set.

5. After baking, remove the pie from direct heat and let it cool for about five minutes. Serve it in a warm condition. In the case of leftovers, enjoy it cold.

Notes: *For those of you who are willing to add flavor to this recipe, you may add half tbsp. of any seasoning mix with the meat.*

Smokey Bacon Meatballs

Total Prep & Cooking Time: 30 minutes

Yields: 8 servings

Nutrition Facts: Calories: 280 | Carbs: 1g | Protein: 13g | Fat: 25g | Fiber: 0g

Ingredients:

- Two garlic cloves (skins peeled)
- Eight bacon slices (crumbled and cooked)
- One pound ground chicken or two chicken breasts
- One egg (properly whisked)
- Two drops of liquid smoke
- One tbsp. of onion powder
- Four tbsps. of olive oil

Method:

1. First, take all the ingredients (except for the oil) and add them to the bowl of the food processor and mix everything.

2. You will be able to form about twenty to twenty-four meatballs from the mixture. These balls will be small in size.

3. Now, take a large-sized frying pan, and then heat the oil. Add the meatballs and fry them until they are browned. It will take about five minutes for each side. If you want them to be perfect, then avoid overcrowding and cook in batches.

Steak au Poivre

Total Prep & Cooking Time: 15 minutes

Yields: 1 serving

Nutrition Facts: Calories: 696 | Carbs: 2g | Protein: 42g | Fat: 58g | Fiber: 0g

Ingredients:

- One fillet of mignon (approximately six ounces)
- One thyme sprig
- One tbsp. of salt
- Two tbsps. each of
  - Ghee
  - Peppercorns
- Two garlic cloves (minced)

Method:

1. After you take the steaks out of the refrigerator, season them nicely with salt and then allow them to sit for about half an hour.

2. Use a mortar and pestle to crush the peppercorns completely on a pan or a flat board.

3. Take the steak, and on both sides of it, press the crushed peppercorns.

4. Place a skillet on the oven and heat it. Add the ghee. After that, sauté the thyme and garlic.

5. When you notice that the ghee has become hot, place the pieces of steak in the pan. Cook each side for about four minutes. The end result will be medium-rare steak.

Skillet Rib Eye Steaks

Total Prep & Cooking Time: 55 minutes

Yields: 2 servings

Nutrition Facts: Calories: 347 | Carbs: 1g | Protein: 22g | Fat: 14.2g | Fiber: 0g

Ingredients:

- Two tsps. of freshly chopped rosemary leaves
- One tsp. of seasoning of your choice
- One tbsp. each of
  o   Olive oil
  o   Unsalted butter
- One rib-eye steak (bone-in)

Method:

1. Take the sheet pan and on it, place the rib-eye steak. Use the seasoning to coat both sides properly. Spread the rosemary leaves on top.

2. Now, keep this steak in the refrigerator for three days after covering. Before cooking, take the steak out and keep it outside at room temperature for half an hour.

3. Place a skillet on the oven and heat it. Add olive oil and butter and wait until all of the butter has melted. Coat the skillet properly with butter by tilting the pan.

4. Now, add the steak to the skillet and cook for about five minutes until you notice that the bottom side has become caramelized and browned. After that, flip it over and baste the other side with oil and butter and cook it for five more minutes.

5. Take the steak off from the pan and slice it into thin pieces after it has cooled down for about five minutes.

Pan-Fried Pork Tenderloin

Total Prep & Cooking Time: 20 minutes

Yields: 2 servings

Nutrition Facts: Calories: 330 | Carbs: 0g | Protein: 47g | Fat: 15g | Fiber: 0g

Ingredients:

- One tbsp. of coconut oil
- To taste – pepper and salt
- One pound of pork tenderloin

Method:

1. Start by cutting the pork tenderloin in two halves.

2. Place your frying pan on the oven on medium flame. Add the oil in the pan and heat it.

3. Once the oil has melted completely, place the two pieces of the pork tenderloin in the oil.

4. Allow the pieces to cook thoroughly. Use tongs to flip the pieces so that all the sides of the pork are evenly cooked.

5. Take a reading on the thermometer, and it should show that the temperature is just below 63 degrees C or 145 degrees F.

6. Allow the pork to cool down after you take it out and then use a sharp knife to cut it into small pieces.

Carnivore Chicken Enchiladas

Total Prep & Cooking Time: 30 minutes

Yields: 10 servings

Nutrition Facts: Calories: 271 | Carbs: 5g | Protein: 25g | Fat: 7g | Fiber: 1.5g

Ingredients:

- Two chicken breasts (skinless, boneless)
- Three tbsps. of bottles lime juice + juice of one fresh lime
- One tsp. of dried garlic
- 16 oz. of sliced chicken
- Chimichurri sauce
- One jar of enchilada sauce
- One bell pepper (thinly sliced)
- Eight oz. each of
  - Cooked spinach
  - Shredded cheese

Method:

For making the shredded chicken,

1. First, take a crockpot and add the shredded pieces of chicken in it. Add the lime juice too.

2. Sprinkle the Chimichurri sauce on top of the chicken and then sprinkle the garlic on top.

3. Now, cook the chicken for about 4-5 hours if you want to cook it on high. Alternatively, if you're going to cook it on low, then set it for 8 hours.

4. Once it is done, use a fork to shred the chicken.

Assembling the enchiladas,

1. Set the temperature of your oven to 400 degrees F and preheat.

2. Take all the other ingredients like pepper and spinach and prep them.

3. The enchilada wrapped will be made by the four slices of chicken.

4. In the middle of the wrapper, add the shredded chicken.

5. Then, on either side, add the cooked spinach, pepper, and some of the cheese.

6. Roll the wrappers carefully and make sure they are firm.

7. Once you have rolled them completely, place them in a pan with the seam sides facing downwards. Then, add the enchilada sauce all over them.

8. Take the remaining portion of the cheese and sprinkle on top of the enchiladas. Bake the preparation for about fifteen minutes in the oven.

9. Serve and enjoy!

PART V

The sirtfood diet is one of the latest diet patterns that has garnered quite the attention. The idea was brought to the market by two nutritionists Glen Matten and Aidan Goggins. The main idea of the diet revolves around sirtuins, which are basically a group of 7 proteins that are responsible for the functioning and regulation of lifespan, inflammation, and metabolism (Sergiy Libert, 2013).

Chapter 1: Health Benefits of the Diet

The benefits are vast. This includes loss in weight, better skin quality, gain in muscle mass in the areas that are very much required, increased metabolic rate, feeling of fullness without having to eat much (this is the power of the foods actually), suppressing the appetite, and leading a better and confident life. This specifically includes an increase in the memory, supporting the body to control blood sugar and blood cholesterol level in a much-advanced way, and wiping out the damage caused by the free radicals and thus preventing them from having adverse impacts on the cells that might lead to other diseases like cancer.

The consumption of these foods, along with the drinks, has a number of observational shreds of evidence that link the sirtfoods with the reducing hazards of several chronic diseases. This diet is notably suited as an anti-aging scheme. Sirtfoods have the ability to satiate the appetite in a natural way and increase the functioning of the muscle. These two points are enough to find a solution that can ultimately help us to achieve a healthy weight. In addition to this, the health-improving impact of these compounds is powerful in comparison to the drugs that are prescribed in order to prevent several chronic diseases like that of diabetes, heart problems, Alzheimer's, etc.

A pilot study was conducted on a total of 39 participants. At the end of the first week, the participants had an increase in muscle mass and also lost 7 pounds on average. Research has proven that in this initial week, the weight loss that is witnessed is mostly from water, glycogen, and muscle, and only one-third of it is from fat (Manfred J. Müller, 2016). The major sirtfoods include red wine, kale, soy, strawberries, matcha green tea, extra virgin olive oil, walnuts, buckwheat, capers, lovage, coffee, dark chocolate, Medjool dates, turmeric, red chicory, parsley, onions, arugula, and blueberries (Kathrin Pallauf, 2013).

Chapter 2: Sirtfood Juice Recipes

Green Juice

Total Prep & Cooking Time: Five minutes

Yields: 1 serving

Nutrition Facts: Calories: 182.3 | Carbs: 42.9g | Protein: 6g | Fat: 1.5g | Fiber: 12.7g

Ingredients:

- Half a green apple
- Two sticks of celery
- Five grams of parsley
- Thirty grams of rocket
- Seventy-five grams of kale
- Half a teaspoon of matcha green tea
- Juice of half a lemon
- One cm of ginger

Method:

1. Juice the kale, rocket, celery sticks, green apple, and parsley in a juicer.

2. Add the lemon juice into the green juice by squeezing it with your hand.

3. Take a glass and pour a little amount of the green juice into it. Add the matcha green tea and stir it in. Then, pour the remaining green juice into the glass and stir to combine everything properly.

4. You can choose to save it for later or drink it straight away.

Blueberry Kale Smoothie

Total Prep & Cooking Time: Five minutes

Yields: 1 serving

Nutrition Facts: Calories: 240 | Carbs: 37.9g | Protein: 17.2g | Fat: 3.6g | Fiber: 7g

Ingredients:

- Half a cup each of
  - Plain low-fat yogurt
  - Blueberries (frozen or fresh)
  - Kale, chopped
- Half a banana
- Half a teaspoon of cinnamon powder
- One tablespoon of flaxseed meal
- One scoop of protein powder
- Half a cup of water (optional)
- Two handfuls of ice (you can add more if you like)

Method:

1. Take a high-speed blender and add all the ingredients in it.

2. Blend everything together until you get a smooth puree.

3. Pour the blueberry kale smoothie in a glass and serve cold.

Tropical Kale Smoothie

Total Prep & Cooking Time: 10 minutes

Yields: 2 servings

Nutrition Facts: Calories: 187 | Carbs: 46.8g | Protein: 3.5g | Fat: 0.5g | Fiber: 4.7g

Ingredients:

- Half a cup to one cup of orange juice (about 120 ml to 240 ml)
- One banana, chopped (use frozen banana, is possible)
- Two cups of pineapple (about 330 grams), chopped (use frozen pineapple if possible)
- One and a half cups of kale (around 90 grams), chopped

Method:

1. Add the chopped bananas, pineapple, kale, and orange juice into a blender and blend everything together until you get a smooth puree.

2. You can add more orange juice if you need to attain a smoothie consistency. The amount of frozen fruit used directly affects the consistency of the smoothie.

3. Pour the smoothie equally into two glasses and serve cold.

Strawberry Oatmeal Smoothie

Total Prep & Cooking Time: 5 minutes

Yields: 2 servings

Nutrition Facts: Calories: 236.1 | Carbs: 44.9g | Protein: 7.6g | Fat: 3.7g | Fiber: 5.9g

Ingredients:

- Half a tsp. of vanilla extract
- Fourteen frozen strawberries
- One banana (cut into chunks)
- Half a cup of rolled oats
- One cup of soy milk
- One and a half tsps. of white sugar

Method:

1. Take a blender. Add the strawberries, banana, oats, and soy milk.

2. Then add sugar and vanilla extract.

3. Blend until the texture becomes smooth.

4. Then pour it into a glass and serve.

Chapter 3: Main Course Recipes for Sirtfood Diet

Green Juice Salad
Total Prep & Cooking Time: Ten minutes

Yields: 1 serving

Nutrition Facts: Calories: 199 | Carbs: 27g | Protein: 10g | Fat: 8.2g | Fiber: 9.2g

Ingredients:

- Six walnuts, halved
- Half of a green apple, sliced
- Two sticks of celery, sliced
- One tablespoon each of
  - Parsley
  - Olive oil
- One handful of rocket
- Two handfuls of kale, sliced
- One cm of ginger, grated
- Juice of half a lemon
- Salt and pepper to taste

Method:

1. To make the dressing, add the olive oil, ginger, lemon juice, salt, and pepper in a jam jar. Shake the jar to combine everything together.

2. Keep the sliced kale in a large bowl and add the dressing over it. Massage the dressing for about a minute to mix it with the kale properly.

3. Lastly, add the remaining ingredients (walnuts, sliced green apple, celery sticks, parsley, and rocket) into the bowl and combine everything thoroughly.

King Prawns and Buckwheat Noodles

Total Prep & Cooking Time: Twenty minutes

Yields: 4 servings

Nutrition Facts: Calories: 496 | Carbs: 53.2g | Protein: 22.2g | Fat: 17.6g | Fiber: 4.8g

Ingredients:

- 600 grams of king prawn
- 300 grams of soba or buckwheat noodles (using 100 percent buckwheat is recommended)
- One bird's eye chili, membranes, and seeds eliminated and finely chopped (and more according to taste)
- Three cloves of garlic, finely chopped or grated
- Three cm of ginger, grated
- 100 grams of green beans, chopped
- 100 grams of kale, roughly chopped
- Two celery sticks, sliced
- One red onion, thinly sliced
- Two tablespoons each of
  - Parsley, finely chopped (or lovage, if you have it)
  - Soy sauce or tamari (and extra for serving)
  - Extra virgin olive oil

Method:

1. Boil the buckwheat noodles for three to five minutes or until they are cooked according to your liking. Drain the water and then rinse the noodles in cold water. Drizzle some olive oil on the top and mix it with the noodles. Keep this mixture aside.

2. Prepare the remaining ingredients while the noodles are boiling.

3. Place a large frying pan or a wok over low heat and add a little olive oil into it. Then add the celery and red onions and fry them for about three minutes so that they get soft.

4. Then add the green beans and kale and increase the heat to medium-high. Fry them for about three minutes.

5. Decrease the heat again and then add the prawns, chili, ginger, and garlic into the pan. Fry for another two to three minutes so that the prawns get hot all the way through.

6. Lastly, add in the buckwheat noodles, soy sauce/tamari, and cook it for another minute so that the noodles get warm again.

7. Sprinkle some chopped parsley on the top as a garnish and serve hot.

Red Onion Dhal and Buckwheat

Total Prep & Cooking Time: Thirty minutes

Yields: 4 servings

Nutrition Facts: Calories: 154 | Carbs: 9g | Protein: 19g | Fat: 2g | Fiber: 12g

Ingredients:

- 160 grams of buckwheat or brown rice
- 100 grams of kale (spinach would also be a good alternative)
- 200 ml of water
- 400 ml of coconut milk
- 160 grams of red lentils
- Two teaspoons each of
  - Garam masala
  - Turmeric
- One bird's eye chili, deseeded and finely chopped (plus more if you want it extra hot)
- Two cms of ginger, grated
- Three cloves of garlic, crushed or grated
- One red onion (small), sliced
- One tablespoon of olive oil

Method:

1. Take a large, deep saucepan and add the olive oil in it. Add the sliced onion and cook it on low heat with the lid closed for about five minutes so that they get softened.

2. Add the chili, ginger, and garlic and cook it for another minute.

3. Add a splash of water along with the garam masala and turmeric and cook for another minute.

4. Next add the coconut milk, red lentils along with 200 ml of water. You can do this by filling the can of coconut milk halfway with water and adding it into the saucepan.

5. Combine everything together properly and let it cook over low heat for about twenty minutes. Keep the lid on and keep stirring occasionally. If the dhal starts to stick to the pan, add a little more water to it.

6. Add the kale after twenty minutes and stir properly and put the lid back on. Let it cook for another five minutes. (If you're using spinach instead, cook for an additional one to two minutes)

7. Add the buckwheat in a medium-sized saucepan about fifteen minutes before the curry is cooked.

8. Add lots of boiling water into the buckwheat and boil the water again— Cook for about ten minutes. If you prefer softer buckwheat, you can cook it for a little longer.

9. Drain the buckwheat using a sieve and serve along with the dhal.

Chicken Curry

Total Prep & Cooking Time: 45 minutes

Yields: 4 servings

Nutrition Facts: Calories: 243 | Carbs: 7.5g | Protein: 28g | Fat: 11g | Fiber: 1.5g

Ingredients:

- 200 grams of buckwheat (you can also use basmati rice or brown rice)
- One 400ml tin of coconut milk
- Eight skinless and boneless chicken thighs, sliced into bite-sized chunks (you can also use four chicken breasts)
- One tablespoon of olive oil
- Six cardamom pods (optional)
- One cinnamon stick (optional)
- Two teaspoons each of
  - Ground turmeric
  - Ground cumin
  - Garam masala
- Two cm. of fresh ginger, peeled and coarsely chopped
- Three cloves of garlic, roughly chopped
- One red onion, roughly chopped
- Two tablespoons of freshly chopped coriander (and more for garnishing)

Method:

1. Add the ginger, garlic, and onions in a food processor and blitz to get a paste. You can also use a hand blender to make the paste. If you have neither, just finely chop the three ingredients and continue the following steps.

2. Add the turmeric powder, cumin, and garam masala into the paste and combine them together. Keep the paste aside.

3. Take a wide, deep pan (preferably a non-stick pan) and add one tablespoon of olive oil into it. Heat it over high heat for about a minute and then add the pieces of boneless chicken thighs. Increase the heat and stir-fry the chicken thighs for about two minutes. Then, reduce the heat and add the curry paste. Let the chicken cook in the curry paste for about three minutes and then pour half of the coconut milk (about 200ml) into it. You can also add the cardamom and cinnamon if you're using them.

4. Let it boil for some time and then reduce the heat and let it simmer for thirty minutes. The curry sauce will get thick and delicious.

5. You can add a splash of coconut milk if your curry sauce begins to get dry. You might not need to add extra coconut milk at all, but you can add it if you want a slightly more saucy curry.

6. Prepare your side dishes and other accompaniments (buckwheat or rice) while the curry is cooking.

7. Add the chopped coriander as a garnish when the curry is ready and serve immediately with the buckwheat or rice.

Chickpea Stew With Baked Potatoes

Total Prep & Cooking Time: One hour and ten minutes

Yields: 4 to 6 servings

Nutrition Facts: Calories: 348.3 | Carbs: 41.2g | Protein: 7.2g | Fat: 16.5g | Fiber: 5.3g

Ingredients:

- Two yellow peppers, chopped into bite-sized pieces (you can also use other colored bell peppers)
- Two 400-grams tins each of
  - Chickpeas (you can also use kidney beans) (don't drain the water if you prefer including it)
  - Chopped tomatoes
- Two cm. of ginger, grated
- Four cloves of garlic, crushed or grated
- Two red onions, finely chopped
- Four to six potatoes, prickled all over
- Two tablespoons each of
  - Turmeric
  - Cumin seeds
  - Olive oil
  - Unsweetened cocoa powder (or cacao, if you want)
  - Parsley (and extra for garnishing)
- Half a teaspoon to two teaspoons of chili flakes (you can add according to how hot you like things)
- A splash of water
- Side salad (optional)
- Salt and pepper according to your taste (optional)

Method

1. Preheat your oven to 200 degrees Celsius.

2. In the meantime, prepare all the other ingredients.

3. Place your baking potatoes in the oven when it gets hot enough and allow it to cook for an hour so that they are cooked according to your preference. You can also use your regular method to bake the potatoes if it's different from this method.

4. When the potatoes are cooking in the oven, place a large wide saucepan over low heat and add the olive oil along with the chopped red onion into it. Keep the lid on and let the onions cook for five minutes. The onions should turn soft but shouldn't turn brown.

5. Take the lid off and add the chili, cumin, ginger, and garlic into the saucepan. Let it cook on low heat for another minute and then add the turmeric along with a tiny splash of water and cook it for a further minute. Make sure that the pan does not get too dry.

6. Then, add in the yellow pepper, canned chickpeas (along with the chickpea liquid), cacao or cocoa powder, and chopped tomatoes. Bring the mixture to a boil and then let it simmer on low heat for about forty-five minutes so that the sauce gets thick and unctuous (make sure that it doesn't burn). The stew and the potatoes should complete cooking at roughly the same time.

7. Finally, add some salt and pepper as per your taste along with the parsley and stir them in the stew.

8. You can add the stew on top of the baked potatoes and serve. You can also serve the stew with a simple side salad.

Blueberry Pancakes

Total Prep & Cooking Time: 25 minutes

Yields: 2 servings

Nutrition Facts: Calories: 84| Carbs: 11g | Protein: 2.3g | Fat: 3.5g | Fiber: 0g

Ingredients:

- 225 grams of blueberries
- 150 grams of rolled oats
- Six eggs
- Six bananas
- One-fourth of a teaspoon of salt
- Two teaspoons of baking powder

Method:

1. Add the rolled oats in a high-speed blender and pulse it for about a minute or so to get the oat flour. Before adding the oats to the blender, make sure that it is very dry. Otherwise, your oat flour will turn soggy.

2. Then, add the eggs and bananas along with the salt and baking soda into the blender and blend them together for another two minutes until you get a smooth batter.

3. Take a large bowl and transfer the mixture into it. Then add the blueberries and fold them into the mixture. Let it rest for about ten minutes to allow the baking powder to activate.

4. To make the pancakes, place a frying pan on medium-high heat and add a dollop of butter into it. The butter will help to make your pancakes really crispy and delicious.

5. Add a few spoonfuls of the blueberry pancake batter into the frying pan and cook it until the bottom side turns golden. Once the bottom turns golden, toss the pancake and fry the other side.

6. Serve them hot and enjoy.

Sirtfood Bites

Total Prep & Cooking Time: 1 hour + 15 minutes

Yields: 15-20 bites

Nutrition Facts: Calories: 58.1 | Carbs: 10.1g | Protein: 0.9g | Fat: 2.3g | Fiber: 1.2g

Ingredients:

- One tablespoon each of
  - Extra virgin olive oil
  - Ground turmeric
  - Cocoa powder
- Nine ounces of Medjool dates, pitted (about 250 grams)
- One ounce (about thirty grams) of dark chocolate (85% cocoa solids), break them into pieces (you can also use one-fourth of a cup of cocoa nibs)

- One teaspoon of vanilla extract (you can also take the scraped seeds of one vanilla pod)
- One cup of walnuts (about 120 grams)
- One to two tablespoons of water

Method:

1. Add the chocolate and walnuts in a food processor and blitz them until you get a fine powder.

2. Add the Medjool dates, cocoa powder, ground turmeric, extra-virgin olive oil, and vanilla extract into the food processor and blend them together until the mixture forms a ball. Depending on the consistency of the mixture, you can choose to add or skip the water. Make sure that the mixture is not too sticky.

3. Make bite-sized balls from the mixture using your hands and keep them in the refrigerator in an airtight container. Refrigerate them for at least an hour before consuming them.

4. To get a finish of your liking, you can roll the balls in some more dried coconut or cocoa. You can store the balls in the refrigerator for up to a week.

Flank Steak With Broccoli Cauliflower Gratin

Total Prep & Cooking Time: 55 minutes

Yields: 4 servings

Nutrition Facts: Calories: 839 | Carbs: 8g | Protein: 43g | Fat: 70g | Fiber: 3g

Ingredients:

- Two tablespoons of olive oil
- Twenty ounces of flank steak
- One-fourth teaspoon salt
- Four ounces of divided shredded cheese
- Half cup of heavy whipping cream
- Eight ounces of cauliflower
- Eight ounces of broccoli
- Salt and pepper

For the pepper sauce,

- One tablespoon soy sauce
- One and a half cups of heavy whipping cream
- Half teaspoon ground black pepper

For the garnishing,

- Two tablespoons of freshly chopped parsley

Method:

1. At first, you have to preheat your oven to four hundred degrees Fahrenheit. Then you need to apply butter on a baking dish (eight by eight inches).

2. Then you have to clean and then trim the cauliflower and broccoli. Then you need to cut them into florets, and their stem needs to be sliced.

3. Then you have to boil the broccoli and cauliflower for about five minutes in salted water.

4. After boiling, you need to drain out all the water and keep the vegetables aside. Then you have to take a saucepan over medium heat and add half portion of the shredded cheese, heavy cream, and salt. Then you need to whisk them together until the cheese gets melted. Then you have to add the cauliflower and the broccoli and mix them in.

5. Place the cauliflower and broccoli mixture in a baking dish. Then you have to take the rest half portion of the cheese and add—Bake for about twenty minutes in the oven.

6. Season with salt and pepper on both sides of the meat.

7. Then you have to take a large frying pan over medium-high heat and fry the meat for about four to five minutes on each side.

8. After that, take a cutting board and place the meat on it. Then you have to leave the meat for about ten to fifteen minutes before you start to slice it.

9. Take the frying pan, and in it, you need to pour soy sauce, cream, and pepper. Then you have to bring it to a boil and allow the sauce to simmer until the sauce becomes creamy in texture. Then you need to taste it and then season it with some more salt and pepper according to your taste.

Kale Celery Salad

Total Prep & Cooking Time: 15 minutes

Yields: 4 servings

Nutrition Facts: 196 | Carbs: 20g | Protein: 5.7g | Fat: 11.5g | Fiber: 4.8g

Ingredients:

- Half a cup of crumbled feta cheese
- Half a cup of chopped and toasted walnuts
- One wedge lemon
- One red apple, crisp
- Two celery stalks
- Eight dates, pitted dried
- Four cups of washed and dried baby kale (stemmed)

For the dressing,

- Three tbsps. olive oil
- One tsp. maple syrup (or you can use any other sweetener as per your preference)
- Four tsps. balsamic vinegar
- Freshly ground salt and black pepper

Method:

1. At first, you have to take a platter or a wide serving bowl. Then you need to place the baby kale in it.

2. Cut the dates into very thin slices, lengthwise. Then you need to place it in another small bowl.

3. After that, you have to peel the celery and then cut them into halves, lengthwise.

4. Then you need to take your knife, hold it in a diagonal angle, and then cut the celery into thin pieces (approximately one to two inches each). Add these pieces to the dates.

5. Then you have to cut the sides off the apple. You need to cut very thin slices from those pieces.

6. Over the apple slices, you need to put some lemon juice to prevent them from browning.

7. For preparing the dressing, you have to take a small bowl, add maple syrup, olive oil, and vinegar. Then you need to whisk them together.

8. Once done, you have to season with freshly ground pepper and two pinches of salt.

9. Before serving, you need to take most of the dressing and pour it over the salad. Then you have to toss nicely so that they get combined. Then you need to pour the rest of the portion of the dressing over the dates and celery.

10. On the top, you have to add the date mixture, feta cheese, apple slices, and walnuts.

Buckwheat Stir Fry

Total Prep & Cooking Time: 28 minutes

Yields: 8 servings

Nutrition Facts: Calories: 258 | Carbs: 35.1g | Protein: 6.8g | Fat: 11.9g | Fiber: 2g

Ingredients:

For the buckwheat,

- Three cups of water
- One and a half cups of uncooked roasted buckwheat groats
- Pinch of salt

For the stir fry,

- Half a cup of finely chopped basil
- Half a cup of finely chopped parsley
- One teaspoon salt
- Four tablespoons of divided red palm oil or coconut oil
- Two cups of drained and chopped marinated artichoke hearts
- Four large bell peppers (sliced into strips)
- Four large minced cloves of garlic
- One bunch of finely chopped kale (ribs removed)

Method:

For making the buckwheat,

1. In a medium-sized pot, pour the buckwheat. Then rinse with cold water and drain the water. Repeat this process for about two to three times.

2. Then add three cups of water to it and also add a pinch of salt. Cover the pot and bring it to a boil.

3. Reduce the heat to low and then cook for about fifteen minutes. Keep the lid on and remove the pot from the heat.

4. Leave it for three minutes and then fluff with a fork.

For making the stir fry,

1. At first, you have to take a ceramic non-stick wok and preheat over medium heat. Then you need to add one tablespoon of oil and coat it. Then you have to add garlic and then sauté for about ten seconds. Then you need to add kale and then add one-fourth teaspoon of salt. Then you need to sauté it accompanied by occasional stirring, until it shrinks in half. Then you have to transfer it to a medium-sized bowl.

2. Then again return to the wok, turn the heat on high, and pour one tablespoon of oil. You need to add one-fourth teaspoon salt and pepper. Then you have to sauté it until it turns golden brown in color. Once done, you need to place it in the bowl containing kale.

3. Then you have to reduce the heat to low, and you need to add two tablespoons of oil. Add the cooked buckwheat and stir it nicely so that it gets coated in the oil. Then after turning off the heat, you need to add the kale and peppers, basil, parsley, artichoke hearts, and half teaspoon salt. Gently stir and serve it hot.

Kale Omelet

Total Prep & Cooking Time: 10 minutes

Yields: 1 serving

Nutrition Facts: Calories: 339 | Carbs: 8.6g | Protein: 15g | Fat: 28.1g | Fiber: 4.4g

Ingredients:

- One-fourth sliced avocado
- Pinch of red pepper (crushed)
- One tsp. sunflower seeds (unsalted)
- One tbsp. of freshly chopped cilantro
- One tbsp. lime juice
- One cup of chopped kale
- Two tsps. of extra-virgin olive oil
- One tsp. of low-fat milk
- Two eggs
- Salt

Method:

1. At first, take a small bowl and pour milk. Then you have to add the eggs and salt to it. Beat the mixture thoroughly. Then take a small non-stick skillet over medium heat, and add one tsp. of oil and heat it. Then add the egg mixture and cook for about one to two minutes, until the time you notice that the center is still a bit runny, but the bottom has become set. Then you need to flip the omelet and cook the other side for another thirty seconds until it is set too. One done, transfer the omelet to a plate.

2. Toss the kale with one tsp. of oil, sunflower seeds, cilantro, lime juice, salt, and crushed red pepper in another bowl. Then return to the omelet on the plate and top it off with avocado and the kale salad.

Tuna Rocket Salad

Total Prep & Cooking Time: 20 minutes

Yields: 4 servings

Nutrition Facts: Calories: 321 | Carbs: 20g | Protein: 33g | Fat: 12g | Fiber: 9.5g

Ingredients:

- Twelve leaves of basil (fresh)
- Two bunches of washed and dried rocket (trimmed)
- One and a half tbsps. of olive oil
- Freshly ground black pepper and salt
- Sixty grams of kalamata olives cut into halves, lengthwise (drained pitted)
- One thinly sliced and halved red onion
- Two coarsely chopped ripe tomatoes
- Four hundred grams of rinsed and drained cannellini beans
- Four hundred grams of drained tuna
- 2 cm cubes of one multigrain bread roll

Method:

1. At first, you need to preheat your oven to 200 degrees Celsius.

2. After that, take a baking tray and line it with a foil.

3. Then you have to spread the cubes of bread over the baking tray evenly.

4. Put the baking tray inside the oven and cook it for about ten minutes until it turns golden in color.

5. In the meantime, you have to take a large bowl and add the olives, onions, tomatoes, cannellini beans, and tuna. Then you need to season it with pepper and salt. Add some oil and then toss for smooth combining.

6. Your next step is to add the basil leaves, croutons, and the rocket. Then you need to toss gently to combine. After that, you can divide the salad into the serving bowls and serve.

Turmeric Baked Salmon

Total Prep & Cooking Time: 30 minutes

Yields: 4 servings

Nutrition Facts: Calories: 448 | Carbs: 2g | Protein: 34g | Fat: 33g | Fiber: 0.2g

Ingredients:

- One ripe yellow lemon
- Half a teaspoon of salt
- One teaspoon turmeric
- One tablespoon of dried thyme
- Half a cup of frozen, salted butter (you may require some more for greasing the pan)
- Four fresh one and a half inches thick salmon fillets (skin-on)

Method:

1. At first, you need to preheat your oven to 400 degrees Fahrenheit. Then with a thin layer of butter, you need to grease the bottom of the baking sheet. Rinse the salmon fillets and pat them dry. Then you have to place the salmon fillets on the buttered baking dish keeping the skin side down.

2. Take the lemons and cut them into four round slices. Remove the seeds and then cut each slice into two halves. Then you will have eight pieces.

3. Take a small dish and combine turmeric, dried thyme, and salt. Then you need to mix them well until they are nicely combined. On the top of the salmon fillets, you need to evenly sprinkle the spice mixture.

4. Place two lemon slices over each salmon fillet.

5. After that, you need to grate the cold butter on the top of the salmon fillets evenly. Allow the butter to meltdown and form a delicious sauce.

6. Then you have to cover the pan with parchment or aluminum foil. Put it inside the oven and cook for about fifteen to twenty minutes according to your desire. The cooking time is dependent on the thickness of the salmon fillets. You can check whether it is done or not by cutting into the center.

7. Once done, remove it from the oven and then uncover it. The butter sauce needs to be spooned over from the tray.

8. Top it off with fresh mint and serve.

Chapter 4: One-Week Meal Plan

Day 1

8 AM – Green Juice

12 PM - Blueberry Kale Smoothie

4 PM – Tropical Kale Smoothie

8 PM – Turmeric Baked Salmon

Day 2

8 AM – Tropical Kale Smoothie

12 PM – Green Juice

4 PM – Strawberry Oatmeal Smoothie

8 PM – King Prawns and Buckwheat Noodles

Day 3

8 AM – Strawberry Oatmeal Smoothie

12 PM – Tropical Kale Smoothie

4 PM – Green Juice

8 PM – Buckwheat Stir Fry

Day 4

8 AM – Blueberry Kale Smoothie

12 PM – Green Juice

4 PM – Green Juice Salad

8 PM – Tuna Rocket Salad

Day 5

8 AM – Green Juice

12 PM – Tropical Kale Smoothie

4 PM – Sirtfood Bites

8 PM – Chicken Curry

Day 6

8 AM – Strawberry Oatmeal Smoothie

12 PM – Green Juice

4 PM – Kale Celery Salad

8 PM – Flank Steak with Broccoli Cauliflower Gratin

Day 7

8 AM – Tropical Kale Smoothie

12 PM – Blueberry Kale Smoothie

4 PM – Kale Omelet

8 PM – Chickpea Stew with Baked Potatoes

9 781913 710897